T0213164

Constructing a Consumer-Focused Industry

The old saying 'safe as houses' is being challenged around the world like never before. Over recent decades homeowners have experienced the devastating effects of defects like asbestos, leaky buildings, structural failings, and more recently the combustible cladding crisis. The provision of safe and secure housing is a critical starting point to ensure that social value can be delivered in the built environment. However, some of these dangerous defects have resulted in a lack of security, safety, health, well-being, and social value for households and the wider community. The problems homeowners experience go beyond the substantial financial costs for defect rectification.

Too often there has been a lack of government and industry support to help the housing consumer through these issues or to prevent them from occurring to begin with. It is time for a rethink and restructure of government policy, support, and industry practices to better protect housing consumers and deliver high-quality and sustainable housing that creates social value.

Through evidence-based research and international case studies, this book focuses on the effects that dangerous defects have on the housing consumer. The ongoing construction cladding crisis is used as a primary case study throughout to highlight these implications, with other previous large-scale defect examples, such as leaky buildings and asbestos. Based upon the range of emerging evidence, we propose ideas for policy makers, construction and built environment professionals, owners corporations, and households on how to move forward towards a higher-quality, sustainable, and socially valuable way of residential living.

Government policy has long focused on 'making industry work' through building regulations and standards. It is now time for greater government and industry focus on the consumer to make 'consumer protection work' in the built environment. There is a need to prevent dangerous defects like combustible cladding, better support consumers when defects emerge, and to create buildings for social value rather than minimum standards. Now is the time to build a better future for the end-user.

Dr. David Oswald is a Senior Lecturer and Deputy Program Manager in the School of Property, Construction and Project Management at RMIT University in Australia. He is a journal editor, reviewer, and PhD examiner and has written multiple award-winning academic publications within construction and the built environment. His recent combustible cladding research with Dr. Moore was used in Victorian parliament (Australia) to demonstrate the need for improving homeowner consumer protection.

Dr. Trivess Moore is a Senior Lecturer in the School of Property, Construction and Project Management at RMIT University in Australia. His research relates to housing quality and performance and focuses on the intersection between technical performance, liveability, social impact, and policy. In addition to recent work with Dr. Oswald on combustible cladding, he has been undertaking research looking at retrofit and the circular economy in Australia.

Social Value in the Built Environment
Series Editors: Ani Raiden and Martin Loosemore

Social Value in Construction
Ani Raiden, Martin Loosemore, Christopher Gorse and Andrew King

Social Value in Practice
Ani Raiden and Andrew King

Constructing a Consumer-Focused Industry
Cracks, Cladding, and Crisis in the Residential Construction Sector
David Oswald and Trivess Moore

Constructing a Consumer-Focused Industry

Cracks, Cladding, and Crisis in the
Residential Construction Sector

David Oswald and Trivess Moore

Routledge
Taylor & Francis Group

LONDON AND NEW YORK

Cover credit: @ Getty Images

First published 2022
by Routledge
4 Park Square, Milton Park, Abingdon, Oxon OX14 4RN

and by Routledge
605 Third Avenue, New York, NY 10158

Routledge is an imprint of the Taylor & Francis Group, an informa business

British Library Cataloguing-in-Publication Data
A catalogue record for this book is available from the British Library

Library of Congress Cataloging-in-Publication Data
Names: Oswald, David, author. | Moore, Trivess, author.
Title: Constructing a consumer-focused industry : cracks, cladding and
crisis in the residential construction sector / David Oswald and Trivess
Moore.
Description: Milton Park, Abingdon, Oxon ; New York, NY : Routledge,
2022. | Series: Social value in the built environment | Includes
bibliographical references and index.
Identifiers: LCCN 2021059886 (print) | LCCN 2021059887 (ebook) |
ISBN 9781032009094 (hardback) | ISBN 9781032007311 (paperback) |
ISBN 9781003176336 (ebook)
Subjects: LCSH: Construction industry. | House construction. |
House buying
Classification: LCC HD9715.8.A2 O89 2022 (print) | LCC
HD9715.8.A2 (ebook) | DDC 338.4/7624—dc23/eng/20211230
LC record available at https://lccn.loc.gov/2021059886
LC ebook record available at https://lccn.loc.gov/2021059887

ISBN: 978-1-032-00909-4 (hbk)
ISBN: 978-1-032-00731-1 (pbk)
ISBN: 978-1-003-17633-6 (ebk)

DOI: 10.1201/9781003176336

Typeset in Goudy
by KnowledgeWorks Global Ltd.

Contents

Illustrations

Figures

Tables

Sidebars

Text Box

Case Studies

Acknowledgements

Some of the flammable cladding and COVID-19 research we have presented in this book has drawn upon wider research projects and we would like to acknowledge the work of our colleagues Associate Professor Simon Lockrey (RMIT University) and Professor Emma Baker (The University of Adelaide) for their key contribution to that research. Outputs from these previous projects are referenced throughout the book but further information can be found in the following references:

Oswald, D., Moore, T., and Lockrey, S., *Combustible costs! Financial implications of flammable cladding for homeowners.* International Journal of Housing Policy, 2021. pp. 1–21. DOI: 10.1080/19491247.2021.1893119.

Oswald, D., Moore, T., and Lockrey, S., *Flammable cladding and the effects on homeowner well-being.* Housing Studies, 2021. pp. 1–20. DOI: 10.1080/02673037.2021.1887458.

Oswald, D., Moore, T., and Baker, E., Post pandemic landlord-renter relationships in Australia, AHURI Final Report No. 344. 2020, Melbourne: Australian Housing and Urban Research Institute Limited. https://www.ahuri.edu.au/research/final-reports/344

Oswald, D., Moore, T., and Baker, E., *Exploring the well-being of renters during the COVID-19 pandemic.* International Journal of Housing Policy, 2022. pp. 1–21. DOI: 10.1080/19491247.2022.2037177.

Series Editor Preface

We are pleased to introduce this third book in the Routledge Social Value in the Built Environment book series: *Constructing a Consumer-Focused Industry: Cracks, Cladding, and Crisis in the Residential Construction Sector*. This book is a very timely contribution to this growing series on social value. The Grenfell Tower tragedy in the United Kingdom, the apartment quality crisis in Australia and Hong Kong, and evidence that poor quality buildings kill and maim more people in natural disasters than the events themselves highlight the immense human impact of poor-quality buildings. These impacts tend to follow social stratifications and impact the poorest and most vulnerable in our communities. By bringing an innovative social value lens to this old debate about quality and defects in construction, this book highlights and documents through numerous case studies and examples the human cost of not getting it right. It highlights the responsibility that all built environment practitioners have to the people who will occupy and use their buildings and infrastructure. And as our cities become denser with the inextricable march of urbanisation around the world, this book lays the foundations for a more in-depth debate about the human impacts of what has so far been a largely technical debate.

This book is a welcome and valuable addition to previous volumes in this series. The first book in the series, *Social Value in Construction*, set out the principles and conceptual foundations for social value, considering the legal and moral arguments for social value, and discussed assessing and measuring social value, the most contested and hotly debated aspect in the field. The second book, *Social Value in Practice*, offers a critical springboard for action to consider, create, and deliver social value in the built environment and to connect social value activity to the global Sustainable Development Goals agenda, through examples of good practice in social value from different perspectives, including clients, designers and architects, a city planner, a consultant, and contractors.

This third book in the series, *Constructing a Consumer-Focused Industry: Cracks, Cladding, and Crisis in the Residential Construction Sector*, shows how social value can be created and destroyed within the built environment, with a focus on the consumer in high-density forms of residential living, such as apartment blocks, in particular. The book makes an important contribution to broadening the consideration of defects during the construction phase and to including the consumer as an important stakeholder in the built environment and a greater focus on quality and performance issues that arise during the occupation phase of dwellings. Social value is discussed in relation to residential design, construction, and handover; ethics and fairness of the process when defects arise; sustainability; liveability; affordability; and safety, health, and well-being. This book has significant implications for policy makers, regulators, industry practitioners, and researchers and makes a much-needed contribution to the growing debate about social value in construction. It should be essential reading for any serious student, researcher, and reflective practitioner who is interested in the impact that the construction industry can have on the communities in which it builds.

In future volumes in the Routledge Social Value in the Built Environment series, we look forward to showcasing progress in the growing social value debate. We particularly welcome diverse contributions from academics and practitioners outside the field of construction who advance our current understanding of the social impacts of materials use, design, community involvement and development, urban planning, environmental management practices, human rights, procurement, social enterprises, managing people and labour practices, organisational governance, and fair business practices, to name but a few important topics in this space. The series will address questions of both theory and practice, and it will be broad in scope, reporting new empirical work, ground-breaking approaches, and exposing good and bad practice through real-life case studies. We welcome proposals from thought-leaders and researchers who are interested in social value in all areas relating to the built environment, and we encourage cross-disciplinary co-authorship from researchers, policy makers, and practitioners from around the world. We hope this volume will inspire prospective authors and editors to submit manuscript proposals to the series about their current research and project interests.

Ani Raiden and Martin Loosemore

1 Cracks, cladding, and crisis in the residential sector

1.1 Introduction

Creating a high-quality and sustainable built environment for the consumer and end-user should be a key priority of the construction industry and policymakers. However, this priority is not being met in many locations around the world (see case studies throughout the book). Recent issues around cracks, cladding, and broader construction crisis have emerged in the residential sector highlighting a growing misalignment between what the construction industry is producing and what housing consumers and regulators demand. As our population grows, and as more people migrate from regional areas to cities, there has been an increasing demand on the provision and maintenance of housing. This is occurring at the same time that the residential sector is having to pivot towards delivering high-quality low-carbon outcomes in line with 2050 climate change targets. These competing outcomes (i.e. providing more total housing *and* providing housing that is more sustainable) are set to change the way we design, construct, and use new and existing housing.

In many regions of the world, housing growth has traditionally occurred through cheap low-density housing with some higher density housing closer to city centres. However, there has been a shift occurring towards the provision of more medium-to high-density housing and less detached housing in recognition of a range of challenges (e.g. space, affordability, sustainability). Both low- and high-density forms of residential living can have building quality issues. For example, many low-density homes across the world use asbestos as insulation, and recently high-rise apartment blocks have been found to being plagued with poor quality and defects.

Issues around residential quality and defects are not geographically bound, and the issues discussed in this book are global in nature. For instance, large structural cracking has resulted in buildings being evacuated and even collapsing in both developing and developed countries. Flammable materials have also been used within and on the facades of high-rise buildings in many countries across the world, resulting in building fires with rapid fire spread. These dangerous defects have led to a crisis in the residential sector, where homes have become unsafe, unsellable, and underperforming. The 'solution' is often to push costly rectification back onto the consumer.

DOI: 10.1201/9781003176336-1

This book is part of a series of books that focus on ways social value can be both created and destroyed within the built environment. The main focus of this book (within the social value series) is on the consumer in high-density forms of residential living, such as apartment blocks. We know from the significant research from around the world that the home, and the wider built environment, can bring significant levels of positive and negative well-being, health, financial, and social value for residents [1–6]. Social value is the social impact that is created by organisations within the built environment for the lives of those affected by their activities [7]. It is about going beyond 'minimum' and 'fit for purpose' design and creating socially sensitive infrastructure or architecture that positively contribute to individuals, households, and communities; not only for the present but also into the future. Examples of creating social value in the built environment include:

- designing buildings for improved safety and security,
- designing and building for resiliency (e.g. changing climate, ability to cope during a energy blackout),
- considering green public spaces in urban planning and leveraging spaces between buildings,
- using and upskilling local construction and sustainability businesses,
- creating sustainable transport options,
- engaging with, and within, the local community,
- considering air quality, resource use, and waste to deliver improved sustainability and performance in the design, construction, use, and end of life phases of a building,
- improving occupant health and well-being,
- reducing living and maintenance costs,
- enhancing neighbour and community relationships, and
- delivering high quality and performing housing for all (including low-income or vulnerable households).

These examples are just a few of the many social value considerations that can improve the quality of life for people within the built environment. While we have provided a definition for social value above, there are numerous definitions across the wider literature. This inevitably brings challenges for policymakers, housing consumers, and the building industry in relation to how social value is measured and understanding the impact it can have on the built environment (since any term devoid of a clear consensus on definition can be difficult to measure). It also creates challenges around how to create policy or industry practices which can enhance such outcomes.

Without measurable or tangible benefits, uptake on investment and research into social value has been slow within the construction industry around the world [8]. There has been limited interest in pursuing social value where it is seen that it might compromise traditional measures (cost, time, quality, and safety) of project delivery success [9]. This is despite an increasing push from some governments to initiate a greater focus on social value. For example, in England and Wales, the Public Services (Social Value) Act 2012 was introduced. This Act requires that public service contracts consider the wider value of a project over its entire lifetime, as opposed to traditional procurement based on cost. The Act aims to promote a greater focus of social value to the built environment over other profit-driven motivations. In Victoria, Australia, the state government has moved to include wider metrics of social value to inform public housing policy. These wider social benefits include improved occupant health leading to less trips to the doctors and reduced living costs. This wider social value has been used to set minimum quality and performance requirements for new public housing that goes significantly beyond minimum standards [2, 10] (see Chapter 9 for more details).

Design and construction practices can create significant social value in residential buildings through, for example, having multi-purpose spaces, green spaces, designing social/communal areas, and using high-quality sustainable approaches to heating and cooling. These benefits for social value can be seen at the individual household level, as well as more broadly across developments and communities [2]. However, there is also the potential to destroy social value, with some structures in the built environment not even reaching a 'fit for purpose' outcome, as building defects emerge and it is not possible to use the structure as intended. In conjunction with potentially destroying social value within such buildings, the emergent defects can also be dangerous. Indeed, 'there's no place like home ... until you discover defects' [11].

The recent impacts of dangerous defects seen in different jurisdictions across the global built environment demonstrate that there is a long way to go for the construction industry and policymakers to deliver social value, on top of the traditional success measures of quality, cost, time, and safety. Academic literature on defects has largely been focused on the construction phase, as opposed to the occupancy stage, with little attention to defects that emerge post-handover [6]. We contend in this book that this is a narrow perspective of defects, and the research should be broadened to also provide a greater focus on quality and performance issues that arise during the occupation phase of the dwelling. When issues do emerge in the occupancy stage, there is often: a lack of social value from industry placed on doing rework, a lack of social value placed on building warranty, and a lack of economic resources available for consumers to challenge the fragmented, transient, and illusive supply chain. The consumer is an essential stakeholder within the built environment, yet the attention they deserve in research and practice is often overlooked. This typically means it is too often left to the consumer to deal with costly rectification work to make their dwellings safe and functional again.

This book draws upon our own evidence-based research, existing academic work, as well as policy and industry responses to discuss the current state of understanding about consumer satisfaction within an international residential built environment context. The chapters draw upon evidenced-based research work undertaken in Australia; but each chapter discusses and situates implications within a global context. Through international, topical, and critical cases of cracked buildings, combustible cladding, and other defects, this book focuses on the housing consumer within the built environment. In terms of social value, the building industry has a poor reputation of qualities such as: always being honest and fair to customers, respecting the consumer in the supply chain by trying to achieve high levels of building quality, and taking responsibility for post-handover defects that are under building warranty. In addition, there is significant social value research emerging on the impact residential design has on sustainability, liveability, affordability, health, and well-being [2, 3, 5, 12–14]. A link between poor building quality, reducing social value, and well-being has become increasingly clearer, particularly in the residential sector.

1.2 Cracks, cladding, and crisis

The case for a greater consumer focus within the built environment and the construction industry has grown in recent years following a general decline of building quality. This decline has manifested from a lack of building to, or above, minimum quality and performance standards and a spate of emergent building defects, causing various health, safety, and well-being risks in different locations around the world. These defects have included widespread leaky buildings, flammable cladding, and structural damage.

Cracks are becoming increasingly common in new buildings. Cracking can occur during the construction phase. For example, if tools are dropped by workers or heavy equipment is placed on floors [15]. Cracks can also emerge later in the occupancy stage, such as on plasterwork and walls as the building shifts or settles [16]. In comparison to older buildings, modern buildings are typically taller, slenderer, have thinner walls, and are built at a much faster pace [17] – meaning modern buildings are at higher risk of stress and are more prone to cracking (and other defects). Cracks can indicate there is a problem nearby, and without inspections, maintenance and rectification, the cracking can extend and accumulate until collapse (in the most severe cases). Thus, while many cracks are superficial and are minor defects that cause minimal cause for concern, structural cracking is a dangerous defect which can be costly and challenging to rectify.

Cracking can occur for a variety of reasons including:

- live loads (a variable load, such as heavy human traffic),
- dead loads (a permanent static load, such as the weight of building material),
- strong sunlight exposure,
- rain moisture and wet areas,
- ground movement, and
- hollowness (e.g. tile not bonded with void underneath).

Cracks can be classified as structural or non-structural. Structural cracks typically occur from incorrect design, poor construction, or overloading [17], whereas non-structural cracks are often due to penetration of moisture or thermal variation, which induces cracks in the building materials that do not cause structural weakening [17].

While cracks can form many different patterns, the width of the cracking can be used for classification. In low-rise buildings, cracks less than 2 mm width are typically aesthetic and can be hairline cracks (less than 0.1 mm) or fine cracks (0.2–2 mm). Cracks greater than 2 mm are more likely to indicate structural significance, where moderate cracking can categorised between 2 mm and 5 mm [18]. It is worth noting that the numbers of cracks are also relevant, and if there are several moderate sized cracks, for example, 3 mm, then rectification work will likely be required [18]. Cracks over 5 mm are serious and indicate that there is almost certainly a compromised structure [18]. Cracks between 15 mm and 25 mm are severe and signify extensive damage, where sections of the wall may require replacement, particularly around openings such as doors and windows [18]. The most concerning category is cracks over 25 mm, which are very severe and suggest structural damage that needs major repair work with partial or complete rebuilding [18].

There are many other factors to consider when trying to understand the cracking, such as the type of dwelling, potential leaky pipework, recent nearby construction work, nearby trees, and the age of the building. For example, many rapidly emerging cracks in a one-year-old house would typically be more concerning than a 20-year-old home with cracks that have not changed for a period of time. Also, cracks in an apartment building are likely to be more concerning than those in a single-storey detached dwelling. This is due to the significant loss of life that could be caused if failure does occur. New apartment buildings with rapidly emerging structural cracking can require urgent repair and evacuation, such as the Opal and Mascot Towers in Australia. In some cases, demolition is the deemed solution, for example, if a nearby building has collapsed and there is a concern for similar builds, such as the Space Towers in Colombia (see Case Study 1.1).

Case Study 1.1   Medellin Space apartment tower case study.

In 2013, a concrete pillar gave way in Tower 6 of the Space apartment complex in Medellin (Colombia) [19]. This failure caused a 54-unit, 24-storey apartment building to collapse, killing 12 people, most of who were undertaking repair work to the defect. While the apartment complex had only been completed in earlier 2013, residents of the building had reported concern about emerging cracks in the ceilings and walls in the lower floors of the apartment building. The local authority ordered an evacuation of Tower 6 the day before it collapsed which averted a larger loss of lives. Emergency rectification work started, although the building collapsed before the work was completed.

Immediately after the collapse, the other five towers in the complex were evacuated due to safety concerns. In subsequent evaluations, the remaining towers were also found to have structural failures and deemed unsafe [20]. They underwent a controlled demolition in 2014.

Analysis into what occurred and likely causes identified a number of design and construction issues which emerged both during and post-construction [19]. In particular, there were structural problems found with some of the piles, including differences in pile lengths and variances in the diameter of the enlarged base. There were also questions raised about the stability of the surrounding soil. The building was found to have collapsed due to the significant increase of the vertical load transmitted by some of the defective columns. The vertical load was transferred from defective columns to neighbouring columns, as the defective columns could not carry the vertical load they were designed for, and the neighbouring columns could not carry this extra load. Researchers calculated that the load transfer to neighbouring piles was a 28% increase in vertical load, resulting in the progressive failure of the building [19, 21].

Legal action was taken against a number of the developers, directors, and other key stakeholders over the collapse. In January 2018, a court sentenced several key stakeholders to between 49 and 51 months jail for the wrongful death of one of the victims. This was overturned on appeal with the ruling stating 'the criminal action was extinguished, because they fully compensated the family of Cantor Molina' [20]. The collapse also led to a number of changes to local construction requirements.

Structural cracking is a global issue that has affected buildings and housing consumers in both developed and developing countries. In many cases, there have sadly been loss of life due to devastating structural collapses. In 1987, there was a collapse of a 16-storey residential project, L'Ambiance Plaza (in the United States), where 28 died. Frank J. Heger, who was awarded the 1992 Construction Index Excellence Award [22] for his investigatory research into the collapse, stated: 'the existence of so many life-threatening deficiencies demonstrates a serious failure of U.S. practice for designing and constructing major buildings to protect public safety' [23]. Sadly, the 2021 Miami building collapse (see Chapter 6) indicates that structural issues remain an ongoing issue.

An issue is that cracks are often not visible during the construction phase or soon after the building construction has been completed. This makes them challenging to identify before handover. Inevitably this means that many cracks will be observed, or even occur, at a later point during the occupancy stage. While cracking represents a physical defect, there are many human issues for occupants that manifest from it.

The risk of structural cracking and failure is a serious concern for the public safety of end users and consumers of the built environment. However, it is not the only health and safety issue which is prominent across the built environment. Recently, there have been thousands of buildings around the world that have been identified with flammable materials, particularly flammable cladding, during occupancy.

Cladding, that is flammable, is the latest dangerous defect within the built environment (at the time of writing). Many homeowners across the world have learnt in recent years that their building or dwelling is at risk from flammable cladding. The cladding crisis has been both labelled the 'combustible cladding crisis' and the 'flammable cladding crisis'. Combustible and flammable do have different definitions. In short, combustibility means an ability to burn in air, and flammability means easily burning. However, to describe the cladding crisis, they have both been used, and therefore within this book, both terms combustible cladding crisis and flammable cladding crisis will be applied.

The form of combustible cladding within the crisis has typically been either expanded polystyrene (EPS) or aluminium composite panels (ACPs). While these are the most common types of flammable cladding, there have been other materials scrutinised as potentially dangerous, including extruded polystyrene (XPS) insulation, high-pressure laminate (HPL), polyurethane foam insulation, and biowood, which is composed of 70% reconstituted timber and 30% polyvinyl chloride (PVC). There has even been some debate on whether green walls (i.e. vegetation grown on or near a wall) constitute a flammable cladding.

EPS is a lightweight plastic material commonly used in packaging. ACPs consist of two thin aluminium flat panels (3–6 mm thick) with non-aluminium core that can be flammable (e.g. polyethylene), which has often been used for signage. It is worth noting that not all ACPs are dangerous, as it depends on what material is sandwiched between the aluminium (within the core). A 100% polyethylene (PE) cores were used on the Grenfell Tower, where a fire caused the loss of 72 lives in the United Kingdom (see Chapter 5). Following the Grenfell Tower tragedy, a report into the building industry in the United Kingdom found that the whole system of regulation was not fit for purpose. The Hackitt report [24], which investigated the building industry following the disaster, stated that the current building regulatory framework:

> … fails to hold to account those responsible for building or building product defects when detected, and fails to support building owners who unwittingly inherit responsibility for unresolved defective work.

Similar reflections have been applied to other construction industries, such as in Australia.

Both EPS and ACP materials became attractive building material options in the global construction market due to their cost, ease of installation, thermal properties, and in the case of ACP, it can be aesthetically pleasing too. ACPs were originally thought to being an innovative material choice gained a level of acceptance within the construction industry soon after their introduction to the market in the 1990s. However, the risks that ACPs pose became evident with the deadly 1993 Sun Valley food-processing factory fire in Hereford (Australia) being an early example [25]. Since these early incidents, there have been very high-profile flammable cladding fires that further demonstrate the seriousness of these defects. As well as the Grenfell Tower disaster (United Kingdom) highlighted above, the deadly Shanghai fire in China also incurred significant loss of life. This is an example of how flammable materials, in conjunction with other safety failings, can lead to a devastating outcome for residents (see Case Study 1.2).

Case Study 1.2   Shanghai building fire.

On the 15th of November 2010, a 28-storey high-rise building in Shanghai (China) caught fire. There were 57 deaths officially reported [21, 26], though a further 56 people remain missing [27] and are presumed dead.

Unlicensed welders, who were working on renovating insulation in the building, are believed to have ignited the fire, which then spread up the building [28]. It took five hours before the blaze was under control, requiring more than 60 fire engines, with 100 residents reportedly saved by firefighters [28].

At least eight people were detained following the tragedy for criminal negligence [29]. Families who lost relatives were compensated 960,000 yuan (US$144,470), with approximately two-thirds of that coming from the government and one-third from social donations [30]. Some relatives deemed the financial compensation unacceptable and would not cover the cost to buy a similar apartment to the ones destroyed in the fire [30, 31].

Following subsequent investigations, it was suspected that flammable materials contributed to the rapid spread of the fire [32]. This was not the first fatal fire incident involving combustible cladding and insulation in China, with a CCTV Tower in Beijing killing a firefighter and causing US$125 million damage [33].

Soon after the tragedy, the government announced that the two neighbouring apartment blocks would be rectified, with flammable material that was blamed for the rapid spread of the fire, being removed from these other two blocks [30]. However, the problem is unlikely to be isolated to these few buildings, since it has been reported that flammable materials, including XPS insulation, polyurethane foam insulation, ACPs, and EPS, are still used in China, despite being linked to several disastrous fires [33]. This has led to academic work into high-rise building fire safety in China, calling for limiting the combustibility of insulation materials used in the exterior wall cladding [34].

Crisis is 'a time of great danger, difficulty or doubt when problems must be solved or important decisions must be made' [35]. It is a term that has been widely used when describing the problems that widespread combustible cladding (and other larger-scale building defects) has caused within the built environment. Research into the cladding crisis remains scarce, and with the exception of a few studies [1, 6, 36–38], the crisis is poorly understood in academic and policy terms.

Prior literature related to combustible cladding has typically been within the field of fire safety engineering with investigation into the fire performance of façade systems [39–41]. For example, there has been some emerging research into the risks that combustible cladding poses. One study used numerical simulation to reveal that any initial fire size of greater than 500 kW will result in the ignition of an ACP (with a flammable polyethylene core) [42]. To provide some context, a burning TV set can generate between around 200 and 500 kW [43].

However, in other relevant fields, literature is limited on, for example: how the crisis manifested from the construction process (the field of construction management); and how to manage the crisis (the field of the built environment). Due to this gap, we made a direct call for more research on the topic through a special issue in the highly regarded journal: *Construction Management & Economics* [44]. The call was entitled: Construction defects, danger, disruption and disputes: A systemic view of the construction industry post-Grenfell. The academic papers published in this special issue in late 2021 explored systemic problems with creating and ensuring building quality and safety in the construction sector [45, 46]. Other previous work has raised the concern of regulatory environment that oversees building construction [47], has raised that there are aspects of building regulations that do not work as intended [48], and has proposed that regulators should set clear expectations for when addressing non-compliant issues [49]. The cladding crisis is a case that has provided increasing evidence that the construction practices and minimum quality and performance standards in the national construction codes have been too low as to be fit for purpose [36].

Our research, which we present in Chapters 3–6, has highlighted the implications this cladding crisis has had on affected homeowners. Revealing that as well as safety concerns, the crisis has caused serious financial stress, suffering, and other negative social or well-being outcomes for the consumers of the construction industry, who have potentially invested their life savings into an unsellable defective product. Many affected homeowners also feel isolated, with a lack of community to turn to, as they are unaware of others in a similar situation.

Combustible cladding is one example of crisis in the residential sector. When the residential sector faces crisis, such as with flammable cladding, or other prior examples, such as asbestos or leaky homes, a stronger response and greater support are required for those affected. However, there is a general lack of robust social structures from industry or government bodies, meaning it can be very difficult for homeowners to know who or where to turn to, and even what their legal rights are and the processes they need to follow when post-handover defects arise.

At the heart of the issue is the lack of consumer protection. The owners of properties remain at risk, as there is often nothing inherent within contracts to force the homebuilder to rectify errors, defects, or omissions [50]. The buyer has minimal legislation or legal rights to fall back on, often leaving litigation the only seemingly viable option for homeowners. However, legal action can be costly and time-consuming, considering it involves investigating complex technical issues, several parties, and large volumes of documents [51]. There is also no guarantee of success, meaning the homeowner risks legal costs and long periods living with defects while the legal case unfolds. There is also the risk the developer or builder uses legal loopholes such as dissolving the business to avoid any legal case against them. These are some of the challenges and problems residents can encounter in a crisis. Yet these issues during the occupancy stage of the building have received limited consideration and even less change to policy or industry practice to address these issues and provide better protection for consumers.

This book aims to draw greater attention to defects that emerge in the occupancy stage by highlighting the problems they cause and proposing ideas for next steps. These proposed ideas are not only for consumers but also for policymakers and stakeholders in the broader construction industry and the built environment. This complements previous academic work on defects that has focused on the construction phase (rather than the occupancy stage). While there are some notable exceptions (such as, the Thane building collapse – Case Study 1.3), the occupancy stage is simply when the building is formally recognised as complete and is occupied with residents. The Thane building collapsed during the construction stage, but also while the building was occupied, highlighted a crisis in the construction sector in India. The crisis of illegal construction included bribery, violations, fake professionals, and disregard for health, safety, and well-being.

Case Study 1.3   Thane building collapse: The dangers of defects during occupancy.

On the 4th of April 2013, a building in Thane (India) collapsed, while the eighth floor was being constructed, killing 74 people. The building was under construction at the time of collapse, and thus did not have an occupancy certificate for the 100–150 construction site workers and their families living in the building [52].

Illegal construction practices had been driven from a growing Mumbai population with a high demand for housing [53]. The building was constructed with cheap, poor-quality materials and was designed by a 'fake architect' who had no degree in architecture [54]. This event led to the arrest of 15 people, including the building material supplier and the architect, for the illegal construction work [29, 55]. The builders were charged with culpable homicide and there were many others arrested for bribery in approving the project and overlooking violations [55]. It was reported that bribed municipal officials suggested to the builders to occupy the building swiftly so that it was more difficult for authorities to act on them [21, 56].

The occupancy stage of a building commences after an occupancy certificate has been awarded. Put simply, it is when the building is deemed suitable for occupancy. In the case of the 2013 Thane building collapse, the building was both under construction and occupied after illegal practices. While building collapses may be more prevalent in countries where illegal housing, bribery, and poor construction practices occur, devastating structural failings have also happened in developed countries too.

When buildings become occupied, defects, both minor and major, can provide difficulties and dangers for those residing in the building. It is essential that occupation only occurs when it is safe and suitable and that consumer protection is in place should defects emerge sometime later. The Thane building collapse is a rare example where the building was occupied before the construction phase was completed. Illegal construction practices in India have been widely reported [52, 53], and issues are still apparent in the year of writing (2021) in Thane, with seven fatalities from a collapsed slab in a building in May [57] and five fatalities from a collapsed building in a landslide in July [58].

The Thane building collapse case highlights a range of relevant themes that will be discussed in this book. This includes the dangers of defects when the building is occupied (Chapter 1) and that this stage deserves more attention (see Chapter 2). The vulnerability within the built environment (see Chapter 3), the lack of regard for safety, health, and well-being (see Chapter 4), the lack of corporate social responsibility (see Chapter 5), and the role of government (see Chapters 8 and 9). A summary of this introductory chapter and an overview of the remaining chapters are presented below.

1.3 Summary

The majority of the world is an owner, landlord, or renter of residential property. Many of whom will have experienced building defects. These people are the customers, consumers, and end users of the residential construction industry. They are critical but often overlooked stakeholders within the built environment.

The residential sector within the built environment has seen an increase in consumer demand as cities densify and attempt to cope with the challenges of urban growth, affordability, and sustainability. Along with this rapid building growth in recent decades has been a decline in building quality in many locations, where the focus has been on cost and time outcomes. Overall, there is limited consumer-focused research within the built environment, and arguably it has never been more necessary, as urban densities grow along with consumer demand for the apartment sector. The customer or consumer of the construction industry is often the homeowner. To purchase a house, people can deposit all the money they have, borrow hundreds of thousands, and pay interest for decades. In return, as homeowners, they expect that the product they are purchasing is a house they can call home, where they can happily spend a large proportion of their life (or rent out to others to enjoy). Increasingly, this is not the case and the old saying 'safe as houses' is challenged like at no time in history.

Academically, the book offers a new perspective on defining defects, highlights the need for more focus on post-handover issues and the implications on social and economic outcomes for consumers. The findings shared are cutting-edge and world leading, with little literature previously published on these topics. Practically, the book encourages ways of incorporating greater emphasis on social values, which could include ethics and fairness of the process when defects arise, honesty from builders, value placed on building warranty, responsibility for fixing defects in a timely manner, the support and community within multi-occupancy buildings when issues arise and physical and mental health impacts to individuals. It is hoped this book will be useful for a range of government policymakers, construction and property professionals, consumers, built environment students, and other stakeholders.

This introductory chapter has opened the discussion on the consumer within the built environment and emphasised the need for greater attention. Below there is a brief overview of each chapter within this book. These chapters are based on evidence-based findings from social science research approaches undertaken in Australia for multiple years. This research is largely on the combustible cladding crisis and demonstrates why improvements are needed to protect and support consumers. The use of case studies, similar to the Medellin (Colombia), Shanghai (China), and Thane (India) cases in this chapter, will also continue throughout this book in the following chapters:

Chapter 2 is an overview of the academic research into construction defects, as well as providing case examples in Australia and New Zealand of dangerous defects. A new definition is proposed to act as a foundation for further work into the human aspect of dangerous defects.

Chapter 3 highlights the existence of vulnerability for homeowners within the built environment through case examples of asbestos in Australia and flammable cladding in South Korea. These vulnerabilities discussed include physical, social, organisational, economic, and political.

Chapter 4 demonstrates the implications that defects can have on homeowners, including the loss of social value, safety risks, and significant reductions in their levels of well-being. Case examples on building evacuation in Australia and flammable cladding in the United Kingdom are used to highlight that this human cost (which has received little acknowledgement compared to the financial cost of replacing the defects).

Chapter 5 discusses the emerging state of corporate social responsibility in the built environment, using the high-profile case examples of the Deepwater Horizon disaster in the Gulf of Mexico and the Grenfell Tower disaster in the United Kingdom. The ways in which the construction industry should demonstrate higher levels of corporate social responsibility are presented, such as honouring building warranties, providing honest advice, and not sacrificing safety for financial gain.

Chapter 6 introduces the challenges that groups of owners within multiple occupancy buildings have when responding to dangerous defects, such as flammable cladding. A case study on the Champlain Tower Collapse (the United States of America) demonstrates how a timely response to emerging dangerous defects is of utmost importance. The chapter explores how professional strata managers can help improve outcomes for consumers in multiple occupancy buildings, but that additional government support is needed in times of crisis.

Chapter 7 highlights different housing consumers through discussion on the landlord and tenant relationships in private rental housing. There is explanation on how the COVID-19 crisis has exacerbated weaknesses in housing systems, and how the landlord-tenant relationship can complicate building defect resolution. The unfolding combustible cladding case of leaseholders in the United Kingdom is presented, followed by ideas for better resolution of defects between landlords and tenants, as well as other tensions, such as rental reduction requests, during periods of crisis.

Chapter 8 argues the need for placing the consumer at the centre of policy-thinking to provide more robust consumer protection. A framework is proposed to demonstrate how consumers could be better protected from dangerous defects.

Chapter 9 emphasises that there is a need to go beyond avoiding cracks, cladding, crisis, as well as other defects and disasters, by designing and constructing buildings that create socially valuable homes and neighbourhoods that residents can be proud of. Through research and multiple case studies of high-quality buildings across the United Kingdom and Australia, the chapter demonstrates what can be achieved to deliver high consumer satisfaction and improved social value outcomes.

Chapter 10 provides overarching conclusions on how to create and protect social value in built environment for consumers. This includes government, industry, and consumer considerations for designing and constructing high-quality buildings for residents, as well as ideas for providing more robust consumer protection, should dangerous defects unexpectedly emerge.

References

1. Oswald, D., *Homeowner vulnerability in residential buildings with flammable cladding.* Safety Science, 2021. **136**. https://doi.org/10.1016/j.ssci.2021.105185
2. Moore, T., Y. Strengers, C. Maller, I. Ridley, L. Nicholls, and R. Horne, *Horsham catalyst research and evaluation.* 2016, Melbourne: RMIT University.
3. Horne, R., *Housing sustainability in low carbon cities.* 2018, London: Taylor & Francis Ltd.
4. Montgomery, C., *Happy city: Transforming our lives through urban design.* 2013, UK: Penguin.
5. CABE, *The value of good design. How buildings and spaces create economic and social value.* 2002, London: Commission for Architecture and the Built Environment. https://www.designcouncil.org.uk/sites/default/files/asset/document/the-value-of-good-design.pdf
6. Oswald, D., T. Moore, and S. Lockrey, *Flammable cladding and the effects on homeowner well-being.* Housing Studies, 2021. pp. 1–20.
7. Raiden, A., M. Loosemore, A. King, and C. Gorse, *Social value in construction.* 2019, Oxon: Routledge.
8. Farag, F.F., P. McDermott, and C. Huelin, *The development of an activity zone conceptual framework to improve social value implementation in construction projects using human activity systems.* In: W. Chan and C.J. Neilson, eds. *Proceedings 32nd Annual ARCOM Conference*, 5–7 September 2016. Manchester: Association of Researchers in Construction Management (ARCOM). pp. 975–984.
9. Barraket, J., and M. Loosemore, *Co-creating social value through cross-sector collaboration between social enterprises and the construction industry.* Construction Management & Economics, 2018. **36**(7): pp. 394–408.
10. Moore, T., L. Nicholls, Y. Strengers, C. Maller, and R. Horne, *Benefits and challenges of energy efficient social housing.* Energy Procedia, 2017. **121**: pp. 300–307.
11. Tronquet, M.C., *There's no place like home … until you discover defects: Do prelitigation statutes relating to construction defect cases really protect the needs of homeowners and developers.* Santa Clara Law Review, 2004. **44**(4): pp. 1249–1286.

12. Sherriff, G., P. Martin, and B. Roberts, *Erneley close passive house retrofit: Resident experiences and building performance in retrofit to passive house standard.* 2018, Salford: University of Salford.

13. Pears, A., and T. Moore, *Decarbonising household energy use: The smart meter revolution and beyond.* In: P. Newton et al., eds. Decarbonising the built environment, 2019. Singapore: Springer. pp. 99–115.

14. Baker, E., L.H. Lester, R. Bentley, and A. Beer, *Poor housing quality: Prevalence and health effects.* Journal of Prevention & Intervention in the Community, 2016. **44**(4): pp. 219–232.

15. Chong, W.K., and S.P. Low, *Assessment of defects at construction and occupancy stages.* Journal of Performance of Constructed Facilities, 2005. **19**(4): pp. 283–289.

16. Paton-Cole, V.P., and A. Gurmu. *A review of defects in low-rise residential buildings in the Australian State of Victoria.* In ICEC-PAQS 2018: Grassroots to Concrete Jungle: Dynamics in Built Environment, 2018. Australian Institute of Quantity Surveyors. Sydney, N.S.W.

17. Ajayi, O.O., O.A. Dare-Abel, L. Ogunbowale, and O.P. Chukwuka, *Evaluation and repair of cracks in post occupancy situations.* International Journal of Development Studies, 2020. 3(2): pp. 194–208.

18. Driscoll, R., *Assessment of damage in low-rise buildings, with particular reference to progressive foundation movement.* 1995, Hertfordshire: Building Research Establishment.

19. Yamin, L.E., J.F. Correal, J.C. Reyes, F. Ramirez, R. Rincón, A.I. Hurtado, and J.F. Dorado, *Sudden collapse of the 27-story space building in Medellin, Colombia.* Journal of Performance of Constructed Facilities, 2018. **32**(3): 04018010.

20. Admin, *Supremo Colombia acquits homicide of former collapsed building managers.* 2019, Spain's News 17/11/2021 https://spainsnews.com/supremo-colombia-acquits-homicide-of-former-collapsed-building-managers/

21. Caicedo, B., E. Alonso, C. Mendoza, and J. Alcoverro, *The collapse of space building.* Géotechnique, 2019. **69**(3): pp. 260–273.

22. Bell, G., *Frank Heger.* 2014, Structure Mag 08/2021/10 https://www.structuremag.org/?p=1556

23. Heger, F., *Public-safety issues in collapse of L'Ambiance Plaza.* Journal of Performance of Constructed Facilities, 1991. **5**(2): pp. 92–112.

24. Hackitt, J., *Building a safer future: Independent review of building regulations and fire safety.* 2018, London: Crown Copyright.

25. The Institution of Fire Engineers, *Sun Valley.* 1993, United Kingdom: The Institution of Fire Engineers 09/30/2019 https://www.ife.org.uk/Firefighter-Safety-Incidents/sun-valley-1993/34014

26. China Daily, *Shanghai fire victims' payouts detailed.* 2010, China.org.cn: China 08/06/2021 http://www.china.org.cn/china/2010-11/24/content_21407451.htm

27. Xinhua, *Shanghai high-rise fire death toll rises to 58.* 2010, china.org.cn 08/06/2021 http://www.china.org.cn/china/2010-11/19/content_21377337.htm

28. Yingying, S., and W. Yiyao, *53 killed in Shanghai as fire engulfs high-rise.* 2010, China Daily: China 08/10/2021 https://www.chinadaily.com.cn/china/2010-11/16/content_11553098.htm

29. BBC News, *Eight held after deadly Shanghai high-rise blaze.* 2010, British Broadcasting Corporation 08/10/2021 https://www.bbc.com/news/world-asia-pacific-11762817

30. Minjie, Z., E. Jia, and N. Yinbin, *Payouts for blaze victims.* 2010, Shanghai Daily 08/06/2021 https://archive.shine.cn/metro/Payouts-for-blaze-victims/shdaily.shtml

31. Burke, C., and A. King, *Generating social value through public sector construction procurement: A study of local authorities and SMEs.* In: E. Raiden and Aboagye-Nimo, eds. *Proceedings 31st Annual ARCOM Conference*, 7–9 September 2015. Lincoln. pp. 387–396.

32. Clem, W., *Flammable foam lined Shanghai tower before fire*. 2010, South China Morning Post 08/06/2021 https://www.scmp.com/article/730860/flammable-foam-lined-shanghai-tower-fire

33. Hughes, P., *Flammable cladding in China: A site study 2020*. 2020, Asia Pacific Fire 08/06/2021 https://apfmag.mdmpublishing.com/flammable-cladding-in-china-a-site-study-2020/

34. Peng, L., Z. Ni, and X. Huang, *Review on the fire safety of exterior wall claddings in high-rise*. Procedia Engineering, 2013. **62**: pp. 663–670.

35. Oxford University Press, *Crisis*. 2021, Oxford Learner's Dictionaries 29/11/2021 https://www.oxfordlearnersdictionaries.com/definition/english/crisis_1

36. Law, T., *Beyond minimum: Proposition for building surveyors to exceed the minimum standards of the construction code*. Journal of Legal Affairs and Dispute Resolution in Engineering and Construction, 2021. **13**(2): p. 03721001.

37. Oswald, D., T. Moore, and S. Lockrey, *Combustible costs! Financial implications of flammable cladding for homeowners*. International Journal of Housing Policy, 2021: pp. 1–21. DOI: 10.1080/19491247.2021.1893119.

38. Martin, W., and J. Preece, *Understanding the impacts of the UK 'cladding scandal': Leaseholders' perspectives*. People, Place and Policy, 2021. **15**(1): pp. 46–53.

39. Northe, C., O. Riese, and J. Zehfuß, *Experimental investigations of the fire behaviour of facades with EPS exposed to different fire loads*. In: S. Vallerent, ed. *2nd International Seminar for Fire Safety of Facades*, MATEC Web of Conferences, 46, 02001, May 11–13, 2016. Lund, Sweden.

40. Sassi, S., P. Setti, G. Amaro, L. Mazziotti, G. Paduano, G. Cancelliere, and M. Madeddu, *Fire safety engineering applied to high-rise building facades*. In: S. Vallerent, ed. *2nd International Seminar for Fire Safety of Facades*, MATEC Web of Conferences, 46, 02001-pp. 1–11, May 11–13, 2016. Lund, Sweden. https://pdfs.semanticscholar.org/4f6f/01ac782bf29694ab6037861ede79d7f5613c.pdf

41. McKenna, S., N. Jones, G. Peck, K. Dickens, W. Pawelec, S. Oradei, S. Harris, A. Stec, and R. Hull, *Fire behaviour of modern façade materials – understanding the Grenfell Tower fire*. Journal of Hazardous Materials, 2019. **368**: pp. 115–123.

42. Chen, T., A. Yuen, G. Yeoh, W. Yang, and Q. Chan, *Fire risk assessment of combustible exterior cladding using a collective numerical database*. Fire, 2019. **2**: pp. 1–14.

43. Bengtsson, L., *Enclosure fires*. 2001, Huskvarna, Sweden: NRS Tryckeri.

44. Oswald, D., S. Smith, L.O. Scholtenhuis, and T. Moore, *Construction defects, danger, disruption and disputes: A systemic view of the construction industry post-Grenfell*. Construction Management & Economics, 2021. **39**(12): pp. 949–952.

45. Çıdık, S., and S. Phillips, *Buildings as complex systems: The impact of organisational culture on building safety*. Construction Management & Economics, 2021. **39**(12): pp. 972–987.

46. Brooks, T., J. Gunning, J. Spillane, and J. Cole, *Regulatory decoupling and the effectiveness of the ISO 9001 quality management system in the construction sector in the UK – A case study analysis*. 2021, Construction Management & Economics. pp. 988–1005.

47. Johnston, N., and S. Reid, *An examination of building defects in residential multi-owned properties*. 2019, Melbourne: Deakin University.

48. Meacham, B., *Accommodating innovation in building regulation: Lessons and challenges*. Journal of Risk Research, 2010. **13**(7): pp. 877–893.

49. Smith, M., *Lessons learnt from the Victorian Building Authority's cladding audit in 2015*. Journal of Building Survey, Appraisal & Valuation, 2019. **7**(1): pp. 61–67.

50. Sommerville, J., and J. McCosh, *Defects in new homes: An Analysis of data on 1,696 new UK houses*. Structural Survey, 2006. **24**(1): pp. 6–21.

51. Abdou, A., M. Haggag, and O.A. Khatib, *Use of building defect diagnosis in construction litigation: Case study of a residential building*. Journal of Legal Affairs and Dispute Resolution in Engineering and Construction, 2016. **8**(1): pp. C4515007.

52. BBC News, *India building collapse near Mumbai kills 45.* 2013, British Broadcasting Corporation 03/08/2021 https://www.bbc.com/news/world-asia-india-22036751
53. BBC News, *India ends search for survivors in Mumbai rubble.* 2013, British Broadcasting Corporation 08/03/2021 https://www.bbc.com/news/world-asia-india-22049417
54. PTI, *Fake architect, journalist held in Mumbra building collapse case.* 2013, India TV 03/08/2021 https://www.indiatvnews.com/news/india/fake-architect-journalist-held-in-mumbra-building-collapse-case-21671.html
55. PTI, *Police arrest two more in Thane building collapse case.* 2013, India Today 08/03/2021 https://www.indiatoday.in/india/west/story/police-arrest-thane-building-collapse-shil-phata-custody-municipal-corporation-tragedy-mumbra-158540-2013-04-12
56. Sayed, N., *Mumbra collapse: Police, civic officials took Rs 20 lakh to clear killer building.* 2013, Pune Mirror 08/03/2021 http://www.punemirror.in/article/2/20130408201304080093442796bf37f31a/Mumbra-collapse-Police-civic-officials-took-Rs-20-lakh-to-clear-killer-building.html
57. India TV, *Maharashtra: 7 dead after building collapses in Thane.* 2021, India TV: India 08/03/2021 https://www.indiatvnews.com/news/india/thane-building-collapse-death-toll-707721
58. Asian News International, *5 dead in building collapse following a landslide in Thane.* 2021, NDTV 08/03/2021 https://www.ndtv.com/india-news/5-die-in-landslide-triggered-building-collapse-in-thanes-kalwa-2490116

2 Building defects

Considering the human cost

Key takeaways

- Previous definitions of building defects have centred around a shortcoming of a physical building component, with little consideration on the human implications that defects have for housing consumers.
- A new definition for a 'dangerous defect' is proposed to provide greater focus on the human cost of defects after the construction phase.
- Previous research on defects has typically focused on why they occur and the financial costs of rework for construction organisations during the construction phase.
- There is a need for greater research into the financial and human costs and wider social value implications of defects for the consumers after the construction phase, which has received little policy, industry or academic attention.

Chapter summary

This chapter explores the previous academic literature on building defects. We highlight the various definitions of a 'defect' and how these definitions have failed to engage with the affected consumer. The complexity of defect definitions is exemplified through a case study on the legal battle that occurred after the Lacrosse building fire (Australia), where the defence claimed flammable cladding was used in compliance with construction codes. Following this discussion, we propose a new definition for a 'dangerous defect', which incorporates the potential human costs for consumers. A case study example of a 'dangerous defect' is then provided by discussing the 'Leaky Building' saga (New Zealand). This chapter then explores how most of the focus in academic work has been on the factors causing building defects and the financial costs of rework during the construction phase. This leads to our call for more consumer-focused research on defects, as well as wider building quality/performance and customer satisfaction. We conclude the chapter by stressing that such research-based understanding is essential for providing knowledge that can help with decisions that improve consumer and social value outcomes in the built environment.

DOI: 10.1201/9781003176336-2

2.1 An introduction to defects

Construction defects have been discussed within academic research for residential buildings in many regions of the world. For instance, studies in Europe have explored the sources and origins of defects, with Spanish research highlighting common defects included structural stability, which mostly originated from poor workmanship [1]. In Asia, there has been research into defect risks for residents, with analysis finding risks centred around electrical and plumbing work, as well as reinforced concrete defects, where cracks could manifest causing water damage [2]. In North America, studies have explored defect litigation, with researchers identifying that defect litigation cases are typically related to water damage [3]. In South America, work has explored the classification of post-handover defects in an attempt to have a system where defect information can be analysed and understood [4]. In Oceania, studies have explored the reporting of defects, finding there was poor use of reporting for new residential homes [5]. Despite this academic work internationally, defects remain a significant research challenge, with inconsistent findings into their causation, financial costs, and solutions, or opportunities, for reducing, or eliminating, their emergence. This makes informing changes to policy or to construction practices difficult when evidence remains under researched and uncertain.

There is no universal definition for a building defect. However, most definitions revolve around a 'shortcoming' or 'shortfall' in performance. For example, the Oxford Dictionary defines a building defect as 'a shortcoming or falling short in the performance of a building element' [6]. Academic researchers have proposed alternatives to this definition. Some have provided more detail on such shortcomings, defining a building defect as 'a failing or shortcoming in the function, performance, statutory, or user requirements of a building, and might manifest itself within the structure, fabric, services, or other facilities of the affected building' [5]. Others have stipulated that a defect can only manifest after a building is in use, stating a building defect was: 'a shortfall in performance which manifests itself once the building is operational' [7]. There has also been a linkage of defects and rework in other proposed definitions. For example, a defect is 'a deviation of a severity sufficient to require corrective action' [8]; and a building defect is: 'a physical phenomenon that should be corrected and such activity can be regarded as rework' [9]. Indeed, as well as such links with rework [10–12], the word 'defect' has been often used interchangeably [10] with failure [13], snag [14], and fault [15] in previous academic work.

As the above demonstrates, trying to define a building defect has resulted in a broad range of descriptions. This is problematic, as a lack of consistency in terminology can result in inaccurate measurements, costing, and inappropriate strategies to avoid recurrence [11].

It is worth noting that the severity of the defect is not considered in the above (or many) definitions. Indeed, defect severity has received ambiguous attention in academic literature [11]. Previous work has suggested defects can be minor or major, where the defect becomes major when they are over AU$500 and render: 'the building unsafe, uninhabitable, or unusable for the purposes for which the building was designed or intended' [16]. As we highlighted in Chapter 1, there have also been attempts to define defect severity based upon outcomes, such as the width of a crack. While this helps to distinguish the broad difference between major and minor defects, we argue that there should be a more detailed extension of what exactly a major defect is, what it covers, and where in the building life cycle it can occur.

Firstly, there is a lack of clarity about whether a major defect emerges during the construction phase, the occupancy stage, or can be either. Major defects that occur during the construction phase can be less concerning to consumers than major defects that arise while people reside in the building since visible construction phase defects can be fixed before the occupancy stage (through rework). Secondly, there could also be an indication of the longevity of a major defect, since some major defects could be quickly rectified, while others, such as buildings with asbestos, have been defective for decades. To complement the previous academic work on the definitions on defects, we offer a definition that is more consumer-focused and attempts to capture some of these broader considerations and nuances. We propose that high-profile defects, such as asbestos, widespread leaky buildings, and combustible cladding, are in a league of their own, and should be categorised as a 'dangerous defect'. Within this book, we will apply our definition of a dangerous defect:

A major shortfall in the building performance that emerges after the building is in use, which is not rectified in a timely manner, is costly to address, and poses a continuous risk to the occupants' safety, health or well-being, that can last for years.

Within this definition, we have attempted to capture a greater consumer focus by highlighting that dangerous defects are a type of major building defect that occurs:

- when the building is occupied (and not in the construction phase);
- the defect cannot be, or are not, quickly fixed (potentially due to construction work complexities, disputes in accountability for the defect, or it being a widespread problem); and
- the defect poses both an immediate and long-term risk to the safety, health, or well-being of residents and have a significant financial cost.

It is hoped that this categorisation of such extreme major defects as 'dangerous defects', will provide a greater focus to the human cost. An example of a 'dangerous defect' is highlighted in New Zealand 'Leaky Building' case study (see Case Study 2.1). Currently, most building defect definitions are very broad, do not capture severity, and focus on the physical defective material or structure without acknowledging the implications this can have on people. There needs to be greater attention on the social, safety, security, well-being implications defects have for consumers and end-users.

Case Study 2.1   A dangerous defect: The New Zealand leaky building crisis.

Leaky buildings typically occur when water penetrates and then stays in the roof or wall cavity for some time [17–19]. This results in the building's timber frame remaining wet and allows fungal growth to eat away at the timber frame, causing structural and health risks [17, 20, 21]. There can also be extensive water damage to the building's interior [17].

In New Zealand, thousands of timber frames built between 1994 and 2004 were not weather tight, which had led to widespread decay and structural stability issues [19]. The costs of the leaky home crisis have been conservatively estimated at NZ$47 billion, which is a staggering 20% of the country's GDP [22].

A leaky home is an example of a dangerous defect since it meets our definition as follows [17–19, 21, 22]:

- 'A major shortfall in the building performance that emerges after the building is in use' – The impacted homes in New Zealand did not reveal their defects until a period of time after the dwellings were occupied. In some cases, the defect was only revealed upon inspection once the broader issue was raised and there are reports that some dwellings are still being identified as having issues. The defects were across a spectrum from minor (in some cases and if identified early) to complete structural damage rendering some dwellings uninhabitable and not able to meet their basic function.
- 'Which is not rectified in a timely manner' – It has been estimated that between 42,000 to more than 89,000 dwellings have been impacted, with only 3500 being reported as fixed in a 2009 report and an unknown number of dwellings still needing to undergo rectification work [23]. The rectification work itself could take up to 6 months to complete. New Zealand homeowners are reportedly still uncovering leaky homes more than two decades after the issue of leaky homes was raised in New Zealand.
- 'Is costly to address, and poses a continuous risk to the occupants' safety, health or well-being, that can last for years' – There is a direct cost for rectifying impacted dwellings which for some dwellings has topped more than NZ$300,000 and that there is a negative impact on property values for which have not undergone rectification work [24]. There was also a health and well-being toll on occupants from both the defect (e.g. impact of mould and damp on health) and process of rectification (e.g. the emotional toll in navigating the processes). The incremental health costs of damp/mouldy dwellings in New Zealand (including all housing with damp/mould, not just those specific to leaky homes) was calculated (2006 dollars) to be NZ$26 million per year or $900 per home highlighting the significant and continued impact on occupants and owners of impacted dwellings [19].

One of the main causes was the cladding systems used, but also the lack of detailed drawings, unqualified workers, and changes in government regulations [22].

This is another example of where rectifying building defects have been pushed onto the homeowner. While the New Zealand government developed a financial assistance package where impacted owners could claim up to 25% of their repair costs (with a possible additional 25% covered by participating local councils), the remaining costs were still born by the owners rather than the industry stakeholders who caused the defects.

2.2 Building defects: The case for greater consumer-focus

The growth of new homes around the world has put building quality and performance under pressure, as evidenced by the increase in the quantity and range of defects that have emerged [25]. Building defects can be expensive to fix, stressful to deal with, can affect building quality, performance, and value, and can be dangerous. The most common defects include building thermal performance issues, such as cracks in the façade, leakage caused by rainwater, and the function of the windows and the balcony [26]. Other common defects are related to the stability of the structure and inappropriate installation of facades and roofs [10].

Unfortunately, building defects are not uncommon, with various research reports and government white papers highlighting that large proportions of the residential building stock in locations like the United Kingdom and Australia had been impacted by construction defects [27, 28]. The increase of larger properties, such as high-rise residential apartments, has contributed to these statistics. Research has suggested these types of buildings are more prone to defects, with a study finding a direct relationship between the size of the property and the number of errors, defects, and omissions identified [14].

Previous research on building defects has typically focused on the causes of building defects and the financial costs of building defects during the construction phase. There have been many different causes of defects identified, though with little consistency on where the focus for resolution should be. Research has highlighted how managerial factors influence the occurrence of defects. For example, a study found that latent managerial errors are often hidden behind more obvious operative errors, which can lead to incorrect attributions and ineffective remedial action [29]. Another investigation reaffirmed that improvements at managerial and strategical levels within projects should be the focus, as opposed to the operational level [30]. Yet this is not always the case, with other work explaining high levels of inexperienced workers, as well as long chains of subcontracting, have contributed to the poor quality of dwellings [1].

These inconsistencies suggest that the different contexts, in terms of industry practice, regulatory frameworks, and economic climate, have a role in how defects occur. For example, in Denmark, research found the extent of defects was related to planning of budgetary conditions, time schedules and early, continuous quality control [31]. In Norway, it was found that the main reason for defects is incomplete and poor design [32]. Whereas in Spain, research found that defects had arisen from poor workmanship rather than by the quality of the materials, design, or products [1, 10]. In Italy, researchers concluded there had been a steady increase in trend of opting for poor-quality materials [33]. Such complexity surrounding the causes of defects can be unhelpful for homeowners seeking to rectify issues, as it is unclear who is accountable. This can be demonstrated by the Lacrosse fire (Australia) legal dispute, where it took over five years for the affected owners to receive compensation (see Case Study 2.2).

Case Study 2.2   Lacrosse legal case: Was the cladding a non-compliant defect?

On the 25th of November 2014, a high-rise building in Melbourne (Australia), caught fire from a cigarette on a balcony. The building, known as the Lacrosse Tower, had flammable cladding that fuelled the fire. Fortunately, there were no fatalities, but there was still significant fire damage caused. This led to a legal battle between the owners corporation and those that designed and constructed the building.

During the legal proceedings, there was debate as to whether the Aluminium Composite Panel (ACP), which was used as external cladding, was a non-compliant defect or not. The owners corporation alleged that the building surveyor had breached his duty by issuing a building permit: 'for cladding that did not, to the degree necessary, avoid the spread of fire in the building and thus failed to meet the requirements of CP2(a)(iv) the BCA (Building Code of Australia)' [34]. In response to the allegations of certifying a non-compliant product, the building surveyor stated: 'This was a material fully encapsulated within non-combustible aluminium and which passed all Australian Standards in relation to smoke-developed index and spread-of-flame index ... my understanding was that the product that you're referring to, composite panel containing PE, as a product complies with the BCA evidence of suitability A2.2' [34].

In the case, Judge Woodward sided with the owners corporation's claims and determined that the building surveyor had: 'failed to identify that the "fit for the purpose" requirement in BCA A2.1 operated as an overarching requirement and was not necessarily satisfied by a single item of "evidence of suitability" under A2.2' [34]. Judge Woodward further added that: 'I accept the Owners' submission that the Alucobest panels used as part of the external walls of Lacrosse were combustible within the meaning of the BCA and, more particularly, in accordance with the test prescribed in AS1530.1.349. The polyethylene core has a calorific value of 44 MJ/kg, which is similar to petrol, diesel and propane... the external cladding of the building, including an ACP with a 100% polyethylene core, did not meet the performance requirement in clause CP2(a)(iv) of the BCA, because it did not, to the necessary degree, avoid the spread of fire in the building' [10, 34].

The conclusion of the case was four years after the 2014 fire. There was a convoluted combination of the architect, fire engineer, building surveyor, and builder deemed accountable all to varying percentage levels. The compensation was paid by fire engineer (39%), building surveyor (33%), architect (25%), and builder (3%). They were ordered to pay a total of an initial AU$5.7 million in compensation, subject to appeal [34]. The length of the case and the fact there were multiple accountabilities demonstrate the challenge for attributing accountability in the complex construction supply chain, as well as how the path to building compliance is not always 'black and white' (since it can even be debated in court).

The case could have been a 'watershed moment' for all other buildings in Australia that have such non-compliant cladding. However, Judge Woodward highlighted the decision applies to the specific circumstances of Lacrosse only [35], meaning this outcome does not directly help other residents affected in other similar buildings. The confirmation that the cladding product is non-compliant in court means that a future defence will also likely change from disputing the compliance status, to acknowledgement the cladding is a non-compliant defect, but attributing blame elsewhere in the construction supply chain (e.g. building surveyor blaming architect). It is now clear that the cladding is a non-compliant defect, but where accountability should lie for each individual building affected (of thousands in Australia) is less certain [36].

The causes of building defects across various international studies have been found to be inconsistent (as highlighted above). Not unexpectedly, the various solutions that have been proposed to help reduce the quantity and severity of defects are also inconsistent. For example, researchers have suggested that design strategies should be the focus to successfully prevent triggering defects [37]. Elsewhere, it has been reported that there was potential for supply chain participants to propose solutions by identifying root causes of defects along the supply chain [38]. Another study recommended a proactive approach to defect management through the application of 'plan-do-check-act' theory to find missing or inadequate steps [39]. Further work is required to help the industry find solutions to the rising defects within different construction contexts.

As well as research into the causes and proposed solutions of building defects, academic work has also focused on the financial costs of defects. Typically, this has been viewed by practitioners and academics as a problem in the construction phase, with defects causing rework that can be financially costly [40], as well as reducing safety performance [41]. The increased construction safety risks can manifest from the increased production and cost pressures on the project [42, 43].

The financial costs from defects, and subsequent rework, during the construction phase, have been highlighted as an area where improvements need to be made. It has been reported that rework can contribute up to 52% of the cost growth – which is the difference between original and actual contract value on completion [12]. The costs of defects have also been inconsistent, depending on the construction context and contractors involved. For example, on two Australian construction projects it was found that the cost of rework was 3.1% and 2.4% of the overall project contract value [44]. While analysis of an Australian contractor involved in over 1900 rework events across 346 construction projects found their yearly profit was reduced by 28% [40]. In a Swedish case study, it was reported that the cost of reworks was 4.4% of the construction value, and time required to correct them was 7.1% of the total work time [45]. In Nigeria, it was revealed that rework accounted for 5.1% of the completion cost for new buildings [46].

While some of these percentages may seem low on the surface, given the size of these construction projects, the financial cost of such rework is significant. It is also important to note that rework is typically only recognised if: the client has identified the need for correction; or the contractor is able to make a payment claim against the client, a subcontractor, or supplier for the extra rework [47]. Hence, there are defects that require rework which do not fall into these categories and therefore may not be addressed and corrected before the occupancy stage.

These financial costs of defects that are not corrected, or emerge during the occupancy stage, have received less attention in academic literature. Further, it is not only the financial cost that becomes relevant in the occupancy stage but also the human cost due to the experience homeowners go through when attempting to have defects rectified. As American Scientist W. Edwards Deming put it [48]:

> No one knows the cost of a defective product – don't tell me you do. You know the cost of replacing it, but not the cost of a dissatisfied customer.

There needs to be greater attention on the human cost associated with defects that emerge during the occupancy stage, as well as the financial implications for homeowners.

2.3 Next steps: Further consumer-focused building defect research

The construction industry has had various problems in producing quality in a customer-oriented manner. To enhance quality and performance, there is a clear need for research into improving regulatory frameworks, as well as construction monitoring, supervision, and skills, to improve the quality and performance of residential buildings [49]. The research into building defects has largely focused on the construction phase. This previous literature has investigated the causes of defects, potential solutions, and the financial costs involved in contractor rework during the construction phase. However, such construction and building research are yet to adequately cover the financial implications and wider social value implications for homeowners when post-handover rework is required. Further, within academic work, there has been little attention to the human and social value aspects, as well as the financial costs, that occur when defects emerge post-handover. One of the main challenges is that a comprehensive inspection on construction work can only typically be carried out during the construction phase of a project, which imposes limitations on what building inspectors can detect [50]. As a result, homeowners are often left with a range of defects that they need to manage in the occupancy stage.

While there have been a few recent studies on defects in the occupancy stage (which have typically explored the types of defects that emerge post-handover [51, 52]), the implications this can have on the consumer is less understood. There is a need to improve the way in which the liability of dangerous defects are managed [53], as these severe defects that can cause significant emotional and financial harm to homeowners are of greatest concern [36, 54]. The flammable cladding crisis is an example of a dangerous defect, which has further highlighted there is a need for all parties in the building supply chain to improve [55]. Addressing the flammable cladding issue will, unfortunately, take a considerable amount of time to change current practices, and a significant cultural shift in the industry [56]. Hence, this is a seismic challenge that should begin with focusing on how to provide a more customer-focused construction industry in both practice and research.

The research work into causes and costs of buildings defects in the construction phase has been useful in building a body of knowledge within this area. However, there has been a narrow approach to studying defects, and we suggest that future research must be broadened to provide greater focus on defects that are present during the occupancy stage and the implications of these on wider social value. This can, in some cases, arise years after the construction phase; as there is often a gap between the perceived construction quality upon project completion and the defects found later during occupancy [50]. When defects in the occupancy stage do emerge, there is often:

- a lack of social value from industry placed on doing work they believed was already completed,
- a lack of social value placed on building warranty, and
- a lack of economic resources available for consumers to challenge the fragmented, transient, and illusive supply chain.

The consumer is an essential stakeholder within the built environment, yet the attention and support they deserve in research and practice is often overlooked. They have been often left to deal with dangerous defects, with limited repercussions for the key stakeholders who contributed to the issue.

Research has revealed there is a need for contractors to improve their performance with quality assurance, handover procedures, and material to increase levels of customer satisfaction [57]. Studies have also highlighted that homeowners, as a consumer of construction industry, perceive a lack of customer focus for residential buildings [58]. There is a limited number of studies that have explored how poor consumer focus can be improved, as well as understanding the customers' expectations [59]. Studies have proposed that the customer of the construction industry is more closely integrated into the construction and urban development processes to improve building quality and performance and incorporate their perspectives in the design [60–62]. Academic work has also highlighted that there needs to be a greater understanding of what it means to be a customer-orientated construction firm and that senior managers within these organisations need to facilitate and enable more customer-focused strategies and processes [63]. Research into other industries, such as manufacturing, has found that customer satisfaction levels are not only influenced by price and quality but also customer service – which includes an ability to elicit and resolve complaints effectively [61]. There has been research proposing models to measure levels of customer satisfaction from construction products [64, 65], but the use of such tools is at an early stage of development [66]. The topic of customer satisfaction remains under-researched within construction literature [67].

When considering next steps, there is a need for academic work, policy development, and industry practice to have a greater consumer focus in the construction industry and built environment. Since there is a lack of robust evidence on the construction consumer, there is little insight provided for industry practice and policy to act on that will improve the customer outcomes. The vast majority of studies within construction journals focus on the construction phase, meaning the customer is often not adequately considered. This is a call for greater research evidence that will help to improve how consumer perceptions and outcomes.

2.4 Conclusions

Building defects can have significant implications for homeowners when they emerge during the occupancy stage. Previous definitions of a defect, and research focus on defects, have typically been within the construction phase. This work has been useful for identifying causes of defects – though these have been inconsistent across countries and construction contexts. There have also been findings into the financial costs of defects and the rework that follows during the construction phase. These have also reported mixed financial costs, due in part to the various methodologies and various definitions of defects that were used. While research into defects during the construction phase is warranted, there needs to be complementary work that is more consumer focused. This consumer-focused research could explore, for example, the financial and emotional implications when defects emerge during the occupancy stage; the customer service and satisfaction for construction consumers, and the loss of social value when different types of defects arise.

There is also a need for greater focus on the human costs of defects (that emerge after handover to the new property owners). Defects can arise immediately after the change of hands or can manifest after many years. The financial and human costs of these types of defects are poorly understood and deserve greater attention. It is proposed that a new defect definition should be considered that more clearly includes the occupancy stage, the defect severity, and the human aspect. Rather than add another definition of 'defect' to those already in literature, we propose a definition of a 'dangerous defect' that focuses more on the consumer. We have defined a dangerous defect as: 'A major shortfall in the building performance that emerges after the building is in use, which is not rectified in a timely manner, is costly to address, and poses a continuous risk to the occupants' safety, health or well-being, that can last for years'.

This definition represents a starting block for research into dangerous defects, such as studies into the causes of dangerous defects, the human costs associated with them, and solutions for avoiding their recurrence. These issues will be discussed further in the upcoming chapters on homeowner vulnerability (Chapter 3), homeowner well-being (Chapter 4), the role of construction firms have delivering corporate social responsibility for consumers (Chapter 5), the challenges groups of owners have when rectifying dangerous defects within multiple occupancy developments (Chapter 6), the government response to dangerous defects (Chapter 8) and what good design and outcomes look like for consumers (Chapter 9).

References

1. Forcada, N., M. Macarulla, M. Gangolells, and M. Casals, *Posthandover housing defects: Sources and origins*. Journal of Performance of Constructed Facilities, 2013. **27**(6): pp. 756–765.
2. Lee, S., S. Lee, and J. Kim, *Evaluating the impact of defect risks in residential buildings at the occupancy phase*. Sustainability, 2018. **10**(12): pp. 4466–4479.
3. Brogan, E., W. McConnell, and C. Clevenger, *Emerging patterns in construction defect litigation: Survey of construction cases*. Journal of Legal Affairs and Dispute Resolution in Engineering and Construction, 2018. **10**(4): 3718003.
4. Vásquez-Hernández, A., and L. Botero, *Standardizing system of posthandover defects for the construction sector in Colombia*. Journal of Architectural Engineering, 2019. **25**(2): 5019004.
5. Rotimi, F., J. Tookey, and J. Rotimi, *Evaluating defect reporting in new residential buildings in New Zealand*. Buildings, 2015. **5**(1): pp. 39–55.
6. Oxford University, *The concise Oxford Dictionary*. 1964, Oxford: Oxford University.
7. Atkinson, G., *A century of defects*. Building, 1987. pp. 54–55.
8. Burati, J.L., J.J. Farrington, and W.B. Ledbetter, *Causes of quality deviations in design and construction*. Journal of Construction, Engineering and Management, 1992. **118**(1): pp. 34–39.
9. Park, C.S., D.Y. Lee, O.S. Kwon, and X. Wang, *A framework for proactive construction defect management using BIM, augmented reality and ontology-based data collection template*. Automation in Construction, 2013. **33**: pp. 61–71.
10. Forcada, N., M. Macarulla, M. Gangolells, and M. Casals, *Assessment of construction defects in residential buildings in Spain*. Building Research & Information, 2014. **42**(5): pp. 629–640.
11. Mills, A., P. Love, and P. Williams, *Defect costs in residential construction*. Journal of Construction, Engineering and Management, 2009. **135**(1): pp. 12–16.
12. Love, P., *Influence of project type and procurement method on rework costs in building construction projects*. Journal of Construction Engineering and Management, 2002. **128**(1): pp. 18–29.
13. Davis, K., W. Ledbetter, and J. Burati, *Measuring design and construction quality costs*. Journal of Construction Engineering and Management, 1989. **115**(3): pp. 385–400.
14. Sommerville, J., and J. McCosh, *Defects in new homes: An analysis of data on 1,696 new UK houses*. Structural Survey, 2006. **24**(1): pp. 6–21.
15. Ilozor, B., M. Okoroh, and C. Egbu, *Understanding residential house defects in Australia from the State of Victoria*. Building and Environment, 2004. **39**(3): pp. 327–337.
16. Georgiou, J., P. Love, and J. Smith, *A comparison of defects in houses constructed by owners and registered builders in the Australian State of Victoria*. Structural Survey, 1999(17): pp. 160–169.
17. NZ Government, *Leaky buildings*, 2002, NZ Government: Wellington. https://www.parliament.nz/en/pb/research-papers/document/00PLEcoRP02111/leaky-buildings
18. James, B., M. Rehm, and K. Saville-Smith, *Impacts of leaky homes and leaky building stigma on older homeowners*. Pacific Rim Property Research Journal, 2017. **23**(1): pp. 15–34.
19. Howden-Chapman, P., J. Bennett, and R. Siebers, eds., *Do damp and mould matter?: Health impacts of leaky homes*. 2009, New Zealand: Steele Roberts Aotearoa Limited.

20. Douwes, J., *Indoor dampness: An overview of health effects and exposures.* In *Do damp and mould matter?: Health impacts of leaky homes.* 2009. New Zealand: Steele Roberts Aotearoa Limited 2009. pp. 33–43.

21. Chapman, R., P. Howden-Chapman, and N. Wilson, *Estimating the health cost of leaky homes,* in *Do damp and mould matter?: Health impacts of leaky homes.* 2009. New Zealand: Steele Roberts Aotearoa Limited 2009. pp. 121–134.

22. Dyer, P., *Rottenomics: The story of New Zealand's leaky buildings.* 2019, New Zealand: Bateman Books.

23. Resolution Architecture, *Leaky building syndrome FAQs.* 2020: New Zealand 17/11/2021 https://www.resolutionarch.co.nz/faqs/

24. Renovation Works, *The Leaky Homes Crisis Needs To End.* 2021, Scoop Business 17/11/2021 https://www.scoop.co.nz/stories/BU2106/S00046/the-leaky-homes-crisis-needs-to-end.htm

25. Hopkins, T., S. Lu, P. Rogers, and M. Sexton, *Detecting defects in the UK new-build housing sector: A learning perspective.* Construction Management & Economics, 2016. **34**(1): pp. 35–45.

26. Jonsson, A.Z., and R.H. Gunnelin, *Defects in newly constructed residential buildings: Owners' perspective.* International Journal of Building Pathology and Adaptation, 2019. **37**(2): pp. 163–185.

27. Wilson, W., *New-build housing: Construction defects – issues and solutions (England).* 2020, House of Commons Library, UK Government 07/15/2021 https://researchbriefings.files.parliament.uk/documents/CBP-7665/CBP-7665.pdf

28. Johnston, N., and S. Reid, *An examination of building defects in residential multi-owned properties.* 2019, Griffith University: Brisbane.

29. Atkinson, A., *The pathology of building defects: A human error approach.* Engineering, Construction and Architectural Management, 2002. **9**(1): pp. 53–61.

30. Jingmond, M., and R. Ågren, *Unravelling causes of defects in construction.* Construction Innovation, 2015. **15**(2): pp. 198–218.

31. Schultz, C., K. Jørgensen, S. Bonke, and G. Rasmussen, *Building defects in Danish construction: Project characteristics influencing the occurrence of defects at handover.* Architectural Engineering and Design Management, 2015. **11**(6): pp. 423–439.

32. Shirkavand, I., J. Lohne, and O. Lædre, *Defects at handover in Norwegian construction projects.* Procedia – Social and Behavioral Sciences, 2016. **226**: pp. 3–11.

33. Sassu, M., and A. De Faclo, *Legal disputes and building defects: Data from Tuscany.* Journal of Performance of Constructed Facilities, 2014. **28**(4): 5019004.

34. VCAT, *Owners Corporation No.1 of PS613436T v LU Simon Builders Pty Ltd (Building and Property) [2019] VCAT 286,* 2018, Victorian Civil and Administrative Tribunal: Melbourne. https://www.vcat.vic.gov.au/documents/forms/owners-corporation-no1-ps613436t-owners-corporation-no-2-ps613436t-owners

35. Hanmer, G., *Lacrosse fire ruling sends shudders through building industry consultants and governments.* 2019, The Conversation 07/14/2021 https://theconversation.com/lacrosse-fire-ruling-sends-shudders-through-building-industry-consultants-and-governments-112777

36. Oswald, D., T. Moore, and S. Lockrey, Flammable cladding and the effects on home-owner well-being. Housing Studies, 2021. DOI: 10.1080/02673037.2021.1887458.

37. Chong, W., and S. Low, *Latent building defects: Causes and design strategies to prevent them.* Journal of Performance of Constructed Facilities, 2006. **20**(3): pp. 213–221.

38. Taggart, M., L. Koskela, and J. Rooke, *The role of the supply chain in the elimination and reduction of construction rework and defects: An action research approach.* Construction Management & Economics, 2014. **32**(7–8): pp. 829–842.

39. Lundkvist, R., J. Henrik Meiling, and M. Sandberg, *A proactive plan-do-check-act approach to defect management based on a Swedish construction project.* Construction Management & Economics, 2014. **32**(11): pp. 1051–1065.

40. Love, P., J. Smith, F. Ackermann, Z. Irani, and P. Teo, *The costs of rework: Insights from construction and opportunities for learning.* Production Planning & Control, 2018. **29**(13): pp. 1082–1095.

41. Love, P., P. Teo, F. Ackermann, J. Smith, J. Alexander, E. Palaneeswaran, and J. Morrison, *Reduce rework, improve safety: An empirical inquiry into the precursors to error in construction.* Production Planning & Control, 2018. **29**(5): pp. 353–366.

42. Oswald, D., F. Sherratt, and S. Smith, *Managing production pressures through dangerous informality: A case study.* Engineering, Construction and Architectural Management, 2019. **26**(11): pp. 2581–2596.

43. Oswald, D., D. Ahiaga-Dagbui, F. Sherratt, and S. Smith, *An industry structured for unsafety? An exploration of the cost-safety conundrum in construction project delivery.* Safety Science, 2020. **122:** 104535.

44. Love, P., and H. Li, *Quantifying the causes and costs of rework in construction.* Construction Management & Economics, 2000. **18**(4): pp. 479–490.

45. Josephson, P., B. Larsson, and H. Li, *Illustrative benchmarking rework and rework costs in Swedish construction industry.* Journal of Management in Engineering, 2002. **18**(2): pp. 76–83.

46. Oyewobi, L.O., A.A. Oke, B.O. Ganiyu, A.A. Shittu, R.S. Isa, and L. Nwokobia, *The effect of project types on the occurrence of rework in expanding economy.* Journal of Civil Engineering and Construction Technology, 2011. **2**(6): pp. 119–124.

47. Barber, P., A. Graves, M. Hall, D. Sheath, and C. Tomkins, *Quality failure costs in civil engineering projects.* International Journal of Quality & Reliability Management, 2000. **17**(4/5): pp. 479–492.

48. Deming, W.E., *W. Edwards Deming Quotes.* 2021 14/02/2021 https://www.brainyquote.com/quotes/w_edwards_deming_672644

49. Paton-Cole, V., and A. Aibinu, *Construction defects and disputes in low-rise residential buildings.* Journal of Legal Affairs and Dispute Resolution in Engineering and Construction, 2021. **13**(1): 05020016.

50. Chong, W., and S. Low, *Assessment of defects at construction and occupancy stages.* Journal of Performance of Constructed Facilities, 2005. **19**(4): pp. 283–289.

51. Lee, J., Y. Ahn, and S. Lee, *Post-handover defect risk profile of residential buildings using loss distribution approach.* Journal of Management in Engineering, 2020. **36**(4): 4020021.

52. Yoon, S., S. Seunghyun, and S. Kim, *Design, construction, and curing integrated management of defects in finishing works of apartment buildings.* Sustainability, 2021(13): pp. 5382.

53. Davey, C., J. McDonald, D. Lowe, R. Duff, J. Powell, and J. Powell, *Defects liability management by design.* Building Research & Information, 2011. **36**(2): pp. 145–153.

54. Oswald, D., T. Moore, and S. Lockrey, *Combustible costs! Financial implications of flammable cladding for homeowners.* International Journal of Housing Policy, 2021: pp. 1–21. DOI: 10.1080/19491247.2021.1893119.

55. Hills, R., *Cladding audits: The problem of combustible cladding and the wider problem of NCBPs and non-compliant building work.* Journal of Building Survey, Appraisal & Valuation, 2018. **6**(4): pp. 312–321.

56. Pargeter, A., *Understanding fire safety requirements for high rise buildings.* Journal of Building Survey, Appraisal & Valuation, 2018. **7**(2): pp. 162–172.

57. Kärnä, S., *Analysing customer satisfaction and quality in construction – The case of public and private customers.* Nordic Journal of Surveying and Real Estate Research, 2014. **2**: pp. 67–80.

58. Hopkin, T., S. Lu, P. Rogers, and M. Sexton, *Key stakeholders' perspectives towards UK new-build housing defects.* International Journal of Building Pathology and Adaptation, 2017. **35**(2): pp. 110–123.

59. Forsythe, P., *A conceptual framework for studying customer satisfaction in residential construction.* Construction Management and Economics, 2007. **25**(2): pp. 171–182.

60. Forsythe, P., *Construction service quality and satisfaction for a targeted housing customer.* Engineering, Construction and Architectural Management, 2016. **23**(3): pp. 323–348.
61. Barlow, J., and R. Ozaki, *Achieving 'customer focus' in private housebuilding: Current practice and lessons from other industries.* Housing Studies, 2003. **18**(1): pp. 87–101.
62. Kuronen, M., W. Majamaa, P. Raisbeck, and C. Heywood, *Including prospective tenants and homeowners in the urban development process in Finland.* Journal of Housing and the Built Environment, 2012. **27**(3): pp. 359–372.
63. Dulaimi, M., *The challenge of customer orientation in the construction industry.* Construction Innovation, 2005. **5**: pp. 3–12.
64. Torbica, Ž, and R. Stroh, *Customer satisfaction in home building.* Journal of Construction Engineering and Management, 2001. **127**(1): pp. 82–86.
65. Yang, J.B., and S.C. Peng, *Development of a customer satisfaction evaluation model for construction project management.* Building and Environment, 2008. **43**(4): pp. 458–468.
66. Kärnä, S., V. Sorvala, and J. Junnonen, *Classifying and clustering construction projects by customer satisfaction.* Facilities, 2009. **27**(9/10): pp. 387–398.
67. Othman, A., *An international index for customer satisfaction in the construction industry.* International Journal of Construction Management, 2015. **15**(1): pp. 33–58.

3 Homeowner vulnerability
Dangerous defects

Key takeaways

- Homeowners have been vulnerable to building failures throughout history, including asbestos, leaky buildings, structural collapse, and now flammable cladding.
- Homeowner vulnerability can come in many forms: physical (e.g. dangerous building defect), economic (e.g. financially unable to fix defects), social (e.g. elderly), political (e.g. lack representation), and organisational (e.g. dysfunctional owners corporation).
- In order to create a built environment where homeowners have an opportunity to live and prosper, then vulnerabilities need to be significantly reduced and ideally eliminated.
- Dangerous defects are not 'temporary' problems that are easily fixed. They can last many years (and even decades), meaning there should be greater continuing support and relief for those affected by government and other key stakeholders.

Chapter summary

This chapter discusses homeowner vulnerability in the built environment when dangerous defects emerge. Homeowner vulnerability from dangerous defects have not been adequately considered in practice, research, and government policy. When dangerous defects occur, they can take many years, and sometimes decades, to identify and resolve. This is not only due to the scale of dangerous defects but the complexity and costs involved in addressing the issue. Thus, these dangerous defects are not temporary defects that are quickly or easily fixed, leaving many homeowners vulnerable for long periods of time while often exposed to health and safety issues that result from the defect. This chapter explains the different types of vulnerabilities that homeowners can be exposed to and how such vulnerabilities can destroy any possibility of creating social value within the home.

The chapter begins by introducing the concept of vulnerability, which has previously been used as a theoretical framework for many natural disasters, such

DOI: 10.1201/9781003176336-3

as flooding. Thereafter the devastating effects of asbestos, as a dangerous building material are highlighted in an Australian case study, before introducing combustible cladding as the latest material to cause crisis. Through our research, and a case study in South Korea, we discuss the vulnerability that manifests from combustible cladding. While there are many challenging problems that have become clear from the cladding crisis (and will not be straightforward to solve), there are some next steps proposed for consideration. These steps focus on the greater acknowledgement that dangerous defects are not a temporary problem and ways to consider reducing the greatest long-term vulnerabilities for homeowners.

3.1 Household vulnerability in the built environment

Vulnerability in everyday language is often understood as the possible exposure to something that can harm or attack. It is not necessarily a word that immediately comes to mind when considering homeowners and defects. In fact, informal sayings, such as 'safe as houses', portrays quite the opposite. Homes should always be places of personal safety, and ideally, a valuable financial asset. People place great social value on their past and current homes: they are a place of warmth, safety, security, upbringing, prosperity and are often intertwined with the meaning of home and community. While often housing does provide outcomes around comfort and stability, there should also be recognition that homeownership can also result in, or exacerbate, forms of vulnerability. Thus, while it is important to consider enhancing social value through the built environment, it is also relevant to consider vulnerabilities that threaten creating long-term social value.

Vulnerability has been defined as the potential for losses [1]. For homeowners, potential losses could manifest from floods [2], cyclones [3], earthquakes [4], fires [5], landslides [6], or other natural or man-made hazards. Often the location of the house can help indicate the risk and physical vulnerability that may be expected. Changes to our natural climate are enhancing many of these natural threats, with significant percentages of the housing increasing exposure, risk, and vulnerability resulting from more extreme weather events [7]. For example, over recent decades there has also been an increase in flooding in areas which were typically not considered flood zones (e.g. the Brisbane (Australia) floods in 2010–2011 [8]), highlighting that vulnerability is not always predictable and can change over time.

When considering vulnerability, it should be recognised that there are multiple different forms, which can be related, overlapping, and have been defined in different ways. These vulnerabilities can include:

- *physical vulnerability*, such as dangerous locations or unprotected infrastructure [9];
- *social vulnerability*, which can relate to social demographics [1];
- *political vulnerability*, which could highlight limited access to political power and representation [10];

- *organisational vulnerability*, which could be weak national or local institutions [10]; and
- *economic vulnerability*, which relates to a lack of access to finances or other resources [11].

These types of vulnerabilities can be observed in homeowners and with their properties, for example, within the context of flooding – as a hazard that some households and dwellings are exposed to. These homeowners likely live on a flood plain (physical vulnerability) but may not be able to afford a flood defence barrier (economic vulnerability) and could also be in a more vulnerable social demographic, such as being very elderly or disabled (social vulnerability). The households may also lack the political representation and power to help address flooding threats (political vulnerability) or could be part of a dysfunctional owners corporation in a multiple occupancy building that does not have an ability to help address the threats from floods (organisational vulnerability). The answer to addressing the risks posed by flooding threats is often through insurance. However, increasingly there are reports of insurance companies not covering flood events for high-risk housing or offering unaffordable insurance rates [12]. For such homes, the event of a flood could be financially and emotionally crippling, as well as life-threatening, and can take years or even decades to recover from. This is a clear example of homeownership becoming more of a liability, as opposed to an asset, due to vulnerability.

Households are not only subject to natural disasters, such as flooding, but have been exposed to human-made hazards through building materials and construction practices. Perhaps the most high-profile global example is the use of asbestos in buildings. During the 20th century, asbestos (which is a naturally occurring fibrous silicate mineral) was used in many countries as a building material. It is highly heat-resistant and an excellent electrical insulator. However, asbestos is also harmful when the asbestos fibres are inhaled, eventually leading to cancer or asbestosis (a long-term inflammation and scarring of the lungs). Sadly, it has been estimated that asbestos has caused 255,000 deaths globally [13]. Financially, the loss of income (due to asbestos-related death and disability) has been estimated at US$243,619 per affected household [14]. There have been countless lawsuits requesting financial compensation for such losses, with the eventual total cost of litigation for asbestos in the United States alone being estimated at US$200–265 billion [15].

It was in the early 1900s when studies in the United Kingdom, France, and Italy identified that there were high numbers of lung problems and early deaths in asbestos-mining towns [16]. In 1924, the first death attributed to asbestos was diagnosed in the United Kingdom [17]. However, it was not until the 1970s when countries began acting through restricting the use of asbestos, such as in the United States, Israel, and Sweden [18]. Then in the 1980s, countries including Denmark, Sweden and Iceland went further and began banning all types of asbestos [18]. Despite some countries taking action, there were others, such as Australia, that continued to use asbestos into the new millennium (see Case Study 3.1 below). The tightening of regulations on asbestos has continued to the present day, when even in the year of writing, there is still regulatory action being taken across the globe. For example, on the 1st of June 2021, 17 states in Brazil banned asbestos [18]. The long delay between issue, evidence, and changes to construction practices continues to be an issue with contemporary defect issues that affect household vulnerability.

Case Study 3.1   'Mr Fluffy'.

In the 1960s and 1970s, residential homes in Canberra and New South Wales (both Australia), were insulated with 'asbestosfluf' by a building company, informally known as 'Mr Fluffy'. It is now known that asbestosfluf is one of the most dangerous types of asbestos, since it is a pure form of asbestos that is crushed, easily airborne, and pervasive in its contamination of a building [19]. The Australian Capital Territory (ACT) government (2019) acknowledged that the 'Mr Fluffy' asbestos was: 'particularly dangerous in that it is comprised of loose raw asbestos fibres which were pumped directly into roof spaces' [20].

Despite other countries taking swifter action, it was not until 2003, when there was a national ban on the importation, manufacture, and use of all products containing chrysotile (white) asbestos in Australia [21]. In 2004, the ACT Government Asbestos Taskforce was created, which was later supported by a Community and Expert Reference Group (CERG). This led to 'The Mr Fluffy Legacy Report' [22], which was prepared for the ACT Government Asbestos Removal Taskforce by CERG. The report found that:

> Many Mr Fluffy homeowners found themselves unable to re-purchase or re-build in the same location or wait for the land to be remediated, heightening a sense of displacement. For those who did re-purchase and re-build significant out of pocket costs were incurred. Ongoing health issues are a factor for members of the Mr Fluffy community and access to adequate services is needed, including mental health services, health screening, practitioner training and research.

In 2021, the Australian Federal Government created an AU$16 million government scheme for victims of the loose-fill 'Mr Fluffy' asbestos [23]. This 'Mr Fluffy' fund for residents was put in place for those who have become sick after living in their defective homes [23]. Previous government schemes had only compensated those who installed the insulation and those exposed that had later worked on the homes [23]. Finally, over 50 years on from the first 'Mr Fluffy' homes, the homeowner was being considered, financially, for ill-health. While this was a necessary step in the response, the emotional toll on affected homeowners should also not be ignored. In 'The Mr Fluffy Legacy Report' [22] by the CERG, insight was provided into the devasting experiences of many 'Mr Fluffy' homeowners. For instance, one previous 'Mr Fluffy' homeowner commented: 'I no longer have a home. The ACT Government and asbestos took it from me' [22]. Another stated: 'I found my so-called home could be responsible for the deaths of my children, family, friends and tradespeople' [22].

> The 'Mr Fluffy' form of asbestos was used in over 1000 homes in Canberra between 1968 and 1979 by building company D. Jansen & Co. Pty. Ltd and its successor firms [21]. In the words of Dr. Sue Parker, Senior Australian of the Year 2019:
>
> There should never be another Mr Fluffy.

Asbestos is not the only building defect to have caused significant financial, health, environmental and social devastation. One of the latest examples of a hazardous human-made building material is the flammable cladding that has been used on thousands of buildings across the world, manifesting into 'the cladding crisis'. The word 'cladding' is arguably the new 'asbestos', within the eyes of the general public. It should, however, be noted that while some forms of cladding are highly flammable (such as the cladding on Grenfell Tower), not all forms of cladding are dangerous (see Chapter 1). The dangers posed by combustible cladding have led to a crisis in countries like Australia, the United Kingdom, and elsewhere, which have revealed various homeowner vulnerabilities [24, 25]. This will be the focus for the rest of this chapter.

3.2 Vulnerabilities revealed from defects: The case of combustible cladding

The combustible cladding crisis has highlighted relevant forms of vulnerability, such as physical, social, political, organisational, and economic. When considering physical vulnerability in the built environment, the physical building location has received much focus. Whether the building is on a flood plain, in a bushfire zone, or on a geographic fault line, there clearly is a physical vulnerability that must be considered. Though within the case of combustible cladding, the focus of physical vulnerability is also on the physical building characteristics and building layout, as well as the building location [5]. Upon exposure to a fire, the location (e.g. bushfire zone), building layout (e.g. escape time) and building characteristics (e.g. presence of flammable cladding) show how physical vulnerability is multi-dimensional. It was revealed in our research [5, 24, 25], that flammable cladding on a dwelling in Australia would often lead to the uncovering of further physical vulnerabilities. Once it was recognised that buildings were compromised with flammable cladding a more thorough building inspection would typically be carried out to do a full check into the safety and quality of the building. The scrutinisation of the affected building would inevitably lead to further building problems being identified, some of which were related to fire safety. For example, following a building inspection report, one interview participant explained:

> The fire rails weren't Australian standard … We had the big roll up doors in the building and all the security gates were ready to fall down and kill somebody. The list [of building defects] went on. It was quite extraordinary, and the building defects were very extensive.

The types of fire safety issues in buildings affected by cladding included defective fire doors, egress routes, and sprinklers. For example, financial shortcuts were reported with the use of standard glass panes (instead of fire resistance glass) along the building egress routes [5]. It is clearly important that buildings identified with flammable cladding do not have such fire safety defects that would further compound any risks. This importance has been highlighted on the Torch Tower in Dubai, which has caught fire twice without a fatality. It was reported that the building design and construction enabled firefighters to tackle the blaze and evacuate residents via smoke-free, fire-free safety zones [26]. This example shows how the fire safety of the building is not only about whether flammable cladding is present or not but about the wider fire safety for occupants. Though, where identified, flammable cladding should be removed or rectified on affected buildings, as well as other aspects of the building's fire strategy for robustness. In cases where other significant fire safety issues have been found, residents have had to evacuate, or temporarily move out, as their buildings were deemed unsafe. For example, in December 2020 a ten-storey apartment building in Sheffield (United Kingdom), was identified as not only having issues with flammable cladding but that there were issues with the smoke ventilation systems and fire escape procedures. This led to an emergency prohibition notice being placed on the sixth to the tenth floor, meaning residents on those floors had to move out temporarily while the issue was fixed [27].

In our research, where buildings had fire safety defects alongside flammable cladding, alternative strategies to improve safety outcomes were often put in place by owners corporations, following the advice of government building authorities. The owners corporations, who are a group of owners of a multi-occupancy building (see Chapter 6), have been put in a very challenging position to help try and solve the problems that have emerged. Typically, they do not have the time, expertise, or experience to deal with the challenges that the flammable cladding crisis has posed [5]. This can represent a form of organisational vulnerability, with the owners corporation challenges discussed in more detail in Chapter 6. In our research, it was revealed that the temporary alternative strategies that owners corporations typically implemented included actions such as:

- new fire safety signage, such as exit routes,
- equipment, such as fire extinguishers, and
- rules, such as banning the use of barbeques on balconies.

These temporary measures relied on human behaviour to address physical building failings. Such strategies are always fraught with challenges, as these measures rely on all residents behaving as advised. In our research, homeowners believed the strategies implemented were not effective, since they would witness other residents often breaking the new fire safety rules, such as bans on barbecues. For example, one homeowner stated:

> We smell a barbecue every once in a while, and salivate … One night, there was somebody having a barbecue and it was actually flaming … There were huge flames on the balcony.

It is important that the strategies also need to consider that in many multi-occupancy buildings, there will likely be some occupants who are socially vulnerable, such as the elderly, young and disabled, who may struggle to evacuate in time during an emergency. One respondent in our study stated:

> This is very distressing as I have a young family (1 & 4-year-old kids) living in a building that is a high risk.

In buildings with several fire safety defects, it was clear the physical vulnerabilities went well beyond only the flammable cladding, with other fire safety defects also raising concerns. Similar examples have been reported elsewhere with flammable cladding fires, such as in the Jecheon building fire in South Korea (see Case Study 3.2), which was described as a 'fire trap' due to the many fire safety failings [28].

Case Study 3.2   South Korea combustible cladding fires.

South Korea and other countries such as United Kingdom, UAE, Australia, and China have experienced multiple high-profile façade fires in the last 30 years. In many of these cases, flammable materials have contributed to fuelling and accelerating the fire up the exterior of the building.

On the 1st of October 2010, the Wooshin Golden Suites fire in Busan (South Korea) occurred after a spark from an electric outlet [29]. The super high-rise building was multi-purpose with both residential and commercial units. Combustible cladding propelled a rapid spread of the fire up the exterior to the top of the building within 20 minutes [29]. There were five injuries reported, with some residents having to escape by helicopter [29]. Following the incident, experts commented that such high-rise buildings were defenceless against fires, and this would have resulted in multiple fatalities had the fire been during the night [30].

Sadly, on the 22nd of December 2017, there was a flammable cladding fire in South Korea where 29 people lost their lives. The eight-storey building in Jecheon was a fitness centre with various leisure facilities. It only took seven minutes for the building to be engulfed in flames, with most of those that lost their lives were in the second-floor sauna [28]. Experts described the structure as a fire trap, as alongside the flammable materials, there were other fire safety issues including insufficient emergency exits, the staircases acting like chimneys, the locker rooms designed like a maze, and cars parked illegally blocking emergency vehicle access [28].

Another façade fire event that occurred in South Korea on the 9th October 2020 resulted in the hospitalisation of over 80 people, but thankfully no fatalities were reported [31]. The 33-floor tower block in Ulsan caught fire around 11 pm local time and rapidly spread [31].

The frequency of these fire events in South Korea is also being experienced in many other places in the world. Research has found an alarming trend that façade fires in large buildings have multiplied by seven times worldwide over the last three decades [32]. This highlights the importance of the overall fire safety strategy, particularly for buildings affected by flammable cladding.

The physical characteristics of the building that compromise fire safety are difficult to manage by modifying human behaviour, considering fire safety rule breaking and ignoring fire alarms, are not uncommon occupant behaviors. For example, during a cladding fire at 'The Cube' building in Bolton (United Kingdom), many of the student residents did not evacuate as they thought it was a false alarm [33]. In an attempt to further help combat these human behaviour challenges, 'waking watches' (people who provide 24-hour surveillance and are tasked with prompting people to evacuate the building immediately in the event of fire) have been employed across the United Kingdom [34]. In 2020, three years after the Grenfell disaster, it was reported that there were 430 tower blocks that had 'waking watches' [34]. United Kingdom waking watches have witnessed 300 fires since Grenfell, and their employment has collectively cost nearly £30 million [35]. In our research, we also found that in some high-risk buildings, 24-hour 'fire wardens' were adopting a similar role. For instance, one respondent listed the fire safety changes their buildings had experienced:

> Air conditioners have been disconnected. Council have appointed a fire warden to sit outside property 24 hours a day. Approximately AU\$170,000 spent on installing fire alarm etc. All plants and tan bark have been removed as fire risk. BBQs are banned. Fire window screens have been installed.

These changes can result in a long list of financial costs that are not related to the direct rectification cost of the flammable materials. For example, as well as 24-hour fire wardens/patrols, we found in our research that there were many other different types of costs that emerged for homeowners with flammable cladding including [24]:

- fixing other fire-related defects (e.g. fire doors, sprinklers, fire-resistant glass), as well as the flammable cladding;
- building inspections, reports, and building material tests;
- legal advice and action;
- extra fire trucks to be dispatched when fire alarm sounds; and
- building insurance rises up 250% because of flammable cladding.

These various costs help to explain the economic vulnerability of homeowners, since these costs were additional necessary spending that was on top of the costs to rectify the defective cladding. Hence, while the actual material rectification works were expensive (quoted between AU\$30,000 and AU\$12 million depending on building size and scope [24]), homeowners were also having to pay for the above-listed costs. Many of the homeowners explained that they continued to pay inspections, reports, legal advice, rising building insurance, and other costs, without any building work having commenced on the defective cladding for years. For example, one homeowner stated:

> The process is ridiculously slow, communication is sparse and passive aggressive at best and they are saying it could take 5 to 10 years to be completed.

There were also concerns about house prices and the ability to rent out properties that were affected. For instance, another homeowner explained:

> It devalued my property, made it harder to rent (due to no BBQs/smoking on balconies) and caused a great deal of stress when I was potentially up for a bill of over AU$100,000.

In the United Kingdom, it has been reported that millions of apartments have become unsellable [36]. A spokesperson for the 'End Our Cladding Scandal Campaign' in the United Kingdom stated [37]:

> Many flats affected by this crisis are valued at zero so it's hard for people to raise funds against them ... So there's a likelihood that many people will get themselves into serious debt, face bankruptcy and losing their homes.

The challenges around fixing defects often revolve around cost: who should pay for the defects? This question should depend on the status of the building warranty, which is often in place around six to twelve years, depending on the legal consumer protection frameworks in place across jurisdictions [38, 39]. The case of cladding is already seeing shifts in the new legislation that acknowledges that this protection may not be long enough. For example, following the Grenfell disaster in the United Kingdom, the new Building Safely Bill allows up to 15 years of protection against shoddy construction work [40, 41]. While this is an improvement, it is unlikely to be good enough to solve the cladding crisis. Since an economic problem remains, in that it is costly for individual homeowners to take legal action against builders, who simply dismiss any accountability. For example, one respondent in our research stated:

> There is no communication. We are spending a fortune on the lawyer ... The construction company says: 'it is not their fault'.

The Lacrosse case (see Case Study in Chapter 2) and unfolding Grenfell enquiry (see Case Study in Chapter 5) demonstrate that attributing accountability within the construction supply chain is complex and requires long legal investigations. These costly legal pursuits are typically not possible for individual homeowners and can even be financially crippling for groups of homeowners within an owners corporation (see Chapter 6). Our research found there were participants that were living in an unsafe home for years and unable to sell to escape. For example, one homeowner stated:

> It is an absolute disgrace and complete failure of regulators and Government. The building was constructed in accordance with the building code at the time so to retrospectively change this and create a situation where owners face huge rectification costs is appalling. We cannot sell our apartments until the cladding issue is resolved and we have no idea what that will entail

or cost. Initially we were told our building was so unsafe we may have to be evicted within 7 days but three years later we are still living here, waiting for some arbitrary process to play out with no plan for the future. It is complete bullsh*t.

Such failure to protect consumers within the built environment through building warranty frameworks can be a form of political vulnerability [5]. This is because customers of the construction industry may not receive adequate government support through policy and regulation when it emerges their house becomes defective (and is under warranty).

The case of cladding has highlighted that the consumer can lack political representation and power in the built environment. For example, in Australia there has been only one state (Victoria) that has provided financial assistance (of AU$600 million) to help fix the highest risk homes from cladding (to date). In the other Australian states, where hundreds of buildings have been identified as at risk, New South Wales has only offered interest-free loans [42], while Queensland [43] has so far rejected pleas to cover costs. Even though Victoria has provided financial assistance, it is only going to cover a small number of the highest risk dwellings, with others left to fund rectification work on their own.

This section has highlighted through the case of combustible cladding that there are various forms of vulnerabilities that homeowners can be exposed to. There clearly needs to be an improved way of reducing such vulnerabilities by proactively stopping defects arising in the first place and prioritising their rectification should they be a health, safety, or well-being risk. This is opposed to relying on 'temporary' approaches, such as 24-hour fire warden patrols, that may not adequately reduce risk, can be very costly, may still be needed to be used for years, and still do not address the fundamental issues – the defects.

3.3 Next steps: Acknowledging and reducing vulnerabilities

In current frameworks, it is assumed that if a large defect occurs when a building is under warranty, then the defect will be fixed at no cost to the homeowner. This assumption has simply not been the case for many defects. Across many jurisdictions, warranties offer little protection for dwelling owners within a system which heavily favours builders and developers. Further, the 'temporary' period between the defect being identified, and the defects being addressed, can be many years, even decades. There has been a lack of consideration of the vulnerability during the time it takes to address the defects. This double issue of limited warranty protection, and long delays to fix defects, has been the case for asbestos, leaky buildings, and now combustible cladding. For the next steps, there needs to be an acknowledgement, first and foremost, that when dangerous defects manifest, the process for rectification is different from standard small scale, minor, and one-off defects. Since dangerous defects are significantly more challenging and are a timely issue to address within the current regulatory and industry frameworks. This can have significant implications for the resident's social value, as their safety, health, well-being, and financial security can be at risk for extended periods. This leads to the question of when dangerous defects arise, how can we better support those affected?

Our research suggests that this support can come from reducing the various forms vulnerability: physical, economic, social, organisational, and political. This list of vulnerabilities may not be restricted to these, with there potentially being other forms of vulnerability that emerge following further research (since there has been little previous work within the area of homeowner vulnerability from defects). For example, when considering economic vulnerability, previous research has noted that there are some major financial risks to consider when purchasing a home [44]. However, this academic work has typically focused on the management of mortgages [45–47], despite there being other economic risks to consider [46] (as the case of cladding demonstrates). Indeed, the costs and financial risks from post-handover defects have received little attention in academic literature [24]. Further research is clearly needed to help inform governments and policy-makers for reducing vulnerabilities, and in doing so, improving the likelihood of enhancing the social value of residents.

Perhaps, there are learnings from other industries that can help inform consumer protection improvements within the built environment. For instance, if your new car or mobile phone is defective, you can typically return to the place of purchase and they will fix the problem (quickly), and at no extra cost (if covered by warranty). There are of course many differences between the car industry, the telecommunications industry, and the construction industry, but there still may be opportunities to establish ideas that could be implemented for more robust consumer protection in the built environment. One of the challenges is that the construction industry has complex contracting chains, so if defects do arise, sometimes years after completion, there is a risk that construction organisations blame each other for the defect, leaving the consumer without a clear answer to the question of who will come and address the issue. There needs to be a clearer way to determine accountability specifically for post-handover defects, to speed up the process of rectification and it should not be left to dwelling owners to work out how to do this on their own.

There is an array of problems that are deep and systemic within the way the industry operates, and there is an immense amount of effort and work required to solve these problems. There are both questions about more effectively stopping defects arising in the first place and then supporting consumers affected when they do happen. Our research has shown that when dangerous defects emerge, such as combustible cladding, consumers have little forms of relief, support, and even clear information about the situation. When other extreme events occur, such as flooding, governments often support with forms of disaster relief payments and other assistance. It can be argued that dangerous building defects, which are extreme events, such as the widespread cases of asbestos and combustible cladding, should also be considered for forms of relief and that there is an urgency to provide that relief, support, and information, much like in the example of flooding. This does not necessarily need to follow the same processes, since current disaster relief packages are typically for natural hazards (such as flood events), and high-profile construction defects are human-made. However, such human-made defects still cause harm, and when there is a high-profile crisis that affects thousands of people, it should be considered.

How could such relief for human-made construction defects look? Perhaps, for dangerous defects, the government could finance the rectification of works to solve the defects more efficiently, and in retrospect, pursue legal action against those responsible (see Chapter 8 for more on this). This would aim at addressing physical (e.g. defects), political (e.g. government representation) and economic (e.g. costs of legal action) vulnerabilities that homeowners currently struggle with. It is also about making sure that there is clear and consistent information provided to impacted owners to ensure that lack of knowledge or understanding of the process does not exacerbate the various vulnerabilities.

3.4 Conclusions

Homes should be places of safety and security, where residents can live and prosper. This is not always the case. When dangerous defects arise, such as combustible cladding, homeowners can be vulnerable, in many different forms including:

- physical vulnerability (e.g. cladding and other fire safety defects),
- economic vulnerability (e.g. costs associated with rectification),
- political vulnerability (e.g. lack of robust consumer protection),
- social vulnerability (e.g. an at-risk social demographic in an evacuation), and
- organisational vulnerability (e.g. an owners corporation without the time, expertise, or experience to respond to a building defect crisis).

When dangerous defects arise, the response time to address the issues is very slow. Building failures, including asbestos and the leaky buildings crisis, have shown that the response to rectification can take decades, with the combustible cladding crisis likely to follow a similar path. Critically, this highlights how these issues are not temporary and resolved quickly, rather they are long-term threats to affected homeowners that require greater attention and support.

The problems have manifested from the way the construction industry has been operating and the regulations in place, which continue to support a lack of accountability. The implications are that dangerous defects keep occurring, and with it, any potential social value created within the home can be destroyed. Governments should consider the various forms of vulnerability homeowners are exposed to when a dangerous defect occurs (such as combustible cladding). They should also plan for better support systems to reduce these vulnerabilities, since these dangerous defects are not rectified quickly.

Without more robust consumer protection that more swiftly solves dangerous defects, the consumer will continue to suffer in many different ways, including economically, the loss of social value placed on their home, and their well-being – which is discussed in the following chapter.

References

1. Cutter, S., B. Boruff, and L. Shirely, *Social vulnerability to environmental hazards.* Social Science Quarterly, 2003. **84**(2): pp. 242–261.
2. Linnekamp, F., A. Koedam, and I. Baud, *Household vulnerability to climate change: Examining perceptions of households of flood risks in Georgetown and Paramaribo.* Habitat International, 2011. **35**(3): pp. 447–456.°
3. Hamidi, A., Z. Zeng, and M. Khan, *Household vulnerability to floods and cyclones in Khyber Pakhtunkhwa, Pakistan.* International Journal of Disaster Risk Reduction, 2020. **46**: pp. 447–456.
4. Bolin, R., and L. Stanford, *The Northridge earthquake: Vulnerability and disaster.* 1998, London: Routledge.
5. Oswald, D., *Homeowner vulnerability in residential buildings with flammable cladding.* Safety Science, 2021. **136**(105185); pp. 1–11.
6. Mertens, K., L. Jacobs, J. Maes, J. Poesen, M. Kervyn, and L. Vranken, *Disaster risk reduction among households exposed to landslide hazard: A crucial role for self-efficacy?* Land Use Policy, 2018. **75**: pp. 77–91.
7. Colvin, R., S. Crimp, S. Lewis, and M. Howden, *Implications of climate change for future disasters.* In: A. Lukasiewicz and C. Baldwin, eds. *Natural hazards and disaster justice,* 2020. Singapore: Palgrave Macmillan. pp. 25–48.
8. Lovejoy, J., *The 2010–11 Queensland floods. Causes, impacts and responses to a hydrological hazard.* 2021, ESRI Australia. https://storymaps.arcgis.com/stories/3bf04a8e0892421983cabcee416ee1e5
9. Zakour, M., and D. Gillespie, *Community disaster vulnerability: Theory, research, and practice.* 2012, New York: Springer.
10. Aysan, Y., *Natural disasters: Protecting vulnerable communities.* In *Vulnerability assessment,* 1993. London: Thomas Telford. pp. 1–14.
11. Anderson, M., and P. Woodrow, *Rising from the ashes: Development strategies in times of disaster.* 2019, Oxon: Routledge.
12. Malo, J., *Insurance in parts of the Gold Coast now unavailable or unaffordable due to flood risk.* 2020, Domain. 6/24/2021. https://www.domain.com.au/news/insurance-in-parts-of-the-gold-coast-now-unavailable-or-unaffordable-due-to-flood-risk-934908/
13. Furuya, S., O. Chimed-Ochir, K. Takahashi, A. David, and J. Takala, *Global asbestos disaster.* International Journal of Environmental Research and Public Health, 2018. **15**(5): pp. 1000–1011.
14. Johnson, W., and E. Heler, *The costs of asbestos-associated disease and death.* Health and Society, 1983. **61**(2): pp. 177–194.
15. Carroll, S.J., D.R. Hensler, A. Abrahamse, J. Gross, S. Ashwood, E.M. Sloss, and M. White, *Asbestos litigation costs and compensation.* 2004, Los Angeles: RAND Corporation.
16. Selikoff, I., and D. Lee, *Asbestos and disease.* 1978, New York: Academic Press Inc.
17. Selikoff, I., and M. Greenberg, *A landmark case in asbestosis.* The Journal of the American Medical Association, 1991. **265**(7): pp. 898–901.
18. Kazan-Allen, L., *Chronology of asbestos bans and restrictions.* 2021, International Ban Asbestos Secretariat. http://www.ibasecretariat.org/chron_ban_list.php
19. Greenlight Environmental Services, *Mr Fluffy – one of the worst asbestos contaminations in history.* 2021, Greenlight Environmental Services, 24/06/2021. https://greenlightservices.com.au/blog/mr-fluffy-and-loose-fill-asbestos-insulation-in-victoria/
20. ACT Health, *Health information for households with Mr Fluffy asbestos insulation.* 2019, Canberra: ACT Health. https://www.health.act.gov.au/sites/default/files/2019-05/Asbestos%20-%20Households%20with%20Mr%20Fluffy%20insulation_0.pdf
21. ACT Government, *Historical timeline.* 2021, ACT Government, 24/06/2021. https://www.asbestostaskforce.act.gov.au/about/history/historical-timeline#1968-1979

22. ACT Government, *Mr Fluffy legacy project: Consultation outcomes, report and recommendations.* 2019, Canberra: ACT Government Asbestos Taskforce. https://www.asbestostaskforce.act.gov.au/__data/assets/pdf_file/0007/1563415/mr-fluffy-legacy-project-consultation-outcomes-and-recommendations-report.pdf

23. CareerSpot, *Mr Fluffy fund for residents.* 2021, 24/06/2021. http://federal.governmentcareer.com.au/archived-news/mr-fluffy-fund-for-residents

24. Oswald, D., T. Moore, and S. Lockrey, *Combustible costs! financial implications of flammable cladding for homeowners.* International Journal of Housing Policy, 2021.

25. Oswald, D., T. Moore, and S. Lockrey, *Flammable cladding and the effects on homeowner well-being.* Housing Studies, 2021. pp. 1–20.

26. Chapman, H., *Cladding implicated in new Torch Tower blaze.* 2017, Building Talk, 07/06/2021. https://www.buildingtalk.com/blog-entry/cladding-implicated-in-new-torch-tower-blaze/

27. BBC News, *Sheffield flats evacuated over fire safety failure.* 2020, British Broadcasting Corporation, 03/05/2021. https://www.bbc.com/news/uk-england-south-yorkshire-55286739

28. Chan-Kyong, P., *South Korea blaze evokes Grenfell Tower: experts.* 2017, The Jakarta Post, 07/06/2021. https://www.thejakartapost.com/news/2017/12/22/south-korea-blaze-evokes-grenfell-tower-fire-experts.html

29. White, N., and M. Delichatsios, *Fire hazards of exterior wall assemblies containing combustible components.* 2015, New York: Springer.

30. Mee-yoo, K., *High-rise apartments defenseless against fire.* 2010, The Korea Times, 07/06/2021. http://www.koreatimes.co.kr/www/news/nation/2010/10/113_73908.html

31. BBC News, *Fire at South Korea 33-level tower brought under control.* 2020, British Broadcasting Corporation, 06/07/2021. https://www.bbc.com/news/world-asia-54470378

32. Bonner, M., and G. Rein, *Flammability and multi-objective performance of building façades: Towards optimum design.* International Journal of High-Rise Buildings, 2018. 7(4): pp. 363–374.

33. BBC News, *Bolton flats blaze: Student flats' cladding 'a concern'.* 2019, British Broadcasting Corporation, 06/10/2020. https://www.bbc.com/news/uk-england-manchester-50445311

34. BBC News, *Number of tower blocks in London with 24-hour fire patrol doubles.* 2020, British Broadcasting Corporation, 07/05/2021. https://www.bbc.com/news/uk-england-london-54539618

35. Apps, P., and N. Barker, *Dangerous buildings protected by 'waking watches' have seen 300 fires since Grenfell, new figures reveal.* 2020, Inside Housing, 07/05/2021. https://www.insidehousing.co.uk/news/news/dangerous-buildings-protected-by-waking-watches-have-seen-300-fires-since-grenfell-new-figures-reveal-65904

36. Ducks, C., *The cladding bomb: Why millions of UK homes have become unsellable.* 2021, Yahoo News, 07/06/2021. https://au.news.yahoo.com/uk-homes-property-cladding-house-prices-leaseholders-230152463.html?guccounter=1

37. Tabary, Z., *After Grenfell: British homeowners face bankruptcy to strip unsafe building panels.* 2020, Sight Magazine, 14/11/2020. https://www.sightmagazine.com.au/features/17442-british-homeowners-face-bankruptcy-to-strip-unsafe-building-panels

38. Government of Alberta, *New home warranty – Overview.* 2020, Canada: Government of Alberta, 11/26/2020. https://www.alberta.ca/new-home-warranty-overview.aspx

39. Victoria State Government, *Implied warranties and domestic building insurance.* 2020, Victoria State Government, 11/26/2020. https://www.consumer.vic.gov.au/housing/building-and-renovating/checklists/implied-warranties-and-domestic-building-insurance

40. BBC News, *Grenfell prompts creation of building safety regulator.* 2021, British Broadcasting Corporation, 07/05/2021. https://www.bbc.com/news/business-57716130

41. BBC News, *Homeowners to get 15 years to sue for 'shoddy' workmanship – minister.* 2021, British Broadcasting Corporation, 07/06/2021. https://www.bbc.com/news/uk-politics-57645976

42. O'Sullivan, M., *Interest-free loans part of $1b program to fix NSW's cladding crisis.* 2020, The Sydney Morning Herald, 07/08/2021. https://www.smh.com.au/national/nsw/interest-free-loans-part-of-1b-program-to-fix-nsw-s-cladding-crisis-20201115-p56ep4.html

43. Lynch, L., *State won't pay to fix Queensland cladding crisis.* 2019, Brisbane Times, 07/08/2021. https://www.brisbanetimes.com.au/national/queensland/state-won-t-pay-to-fix-queensland-cladding-crisis-20190717-p527y8.html

44. Sarah, B., and H. Caroline, *The right to buy: Examination of an exercise in allocating, shifting and re-branding risks.* Critical Social Policy, 2013. **33**(1): pp. 17–36.

45. Aalbers, M., *The financialization of housing: A political economy approach.* 2016, New York: Oxon.

46. Smith, S., B. Searle, and N. Cook, *Rethinking the risks of home ownership.* Journal of Social Policy, 2009. **38**(1): pp. 83–102.

47. Waldron, R., and D. Redmond, *'We're just existing, not living!' Mortgage stress and the concealed costs of coping with crisis.* Housing Studies, 2017. **32**(5): pp. 584–612.

4 The real cost?

Well-being implications for the consumer

Key takeaways

- The real cost of defects is not only the potential financial losses but also the emotional experience when trying to fix defects, which has resulted in significant mental distress and other negative health outcomes for homeowners.
- Combustible cladding is the latest dangerous defect that has led to deteriorating mental health and other negative health outcomes of homeowners, due to financial stresses, safety concerns, and challenges in achieving timely defect rectification.
- The built environment plays a significant role in enhancing or reducing well-being and therefore future government policy should have a greater focus on policy from a well-being perspective, as opposed to only an economic perspective.
- Well-being-based policy should be considered not only during design and construction phases but also when managing dangerous defects, such as combustible cladding.

Chapter summary

This chapter explores the real cost of defects. While the economics of defects is important (in terms of how much they will cost to fix and who is responsible for paying for the rectification work), the real cost is much more than only the financial considerations of the direct defect rectification work. Defects result in a reduction of social value and an increase of mental stress, which has implications for the overall well-being of residents, who are often living with defects for extended periods before they are fixed. This emotional toll is part of the real cost and has significant implications for safety, health, and well-being that are not adequately considered in response to housing defects that emerge post-construction.

This chapter will begin by exploring how the built environment can more carefully consider social value in design and construction. Higher levels of social value have a range of safety, health, and well-being benefits for consumers. It will then turn to the opposite side of the scale, where minimum standards are not met and defects arise, which can be detrimental to the safety, health, and well-being

DOI: 10.1201/9781003176336-4

of consumers. This will be explored through the latest high-profile defect: the case of combustible cladding. Lastly, the final remarks explore the policy opportunities for improving the way defects are managed, to improve the well-being of consumers during the rectification processes. It is recommended that there is further attention on policy that is centred on well-being, to complement current policies which largely try to solve the problems of defects through an economic lens.

4.1 An introduction to well-being in the built environment

The built environment is the human-made space where people live, work, and recreate on a day-to-day basis [1]. These human-made spaces are not just for our safety and security but are places for us to live and prosper. Considering that people can spend over 90% of their life within the built environment [2] and that we are becoming an increasingly urbanised population, it is important to carefully consider the social value within this context, to enhance the liveability and well-being through well-designed and constructed buildings and structures [3–7].

Having safe and healthy cities has become more challenging in recent years with increasing population growth. These challenges have prompted the World Health Organisation to encourage countries globally to prioritise health through the built environment via their healthy cities movement [8]. As populations grow in cities, this can result in more noise, greater crowds, a lack of green spaces, and a reduction of safety and, conversely for some people, lead to *more* isolation. These factors can contribute to a range of negative physical and mental health implications, including distress, depression, aggression, and violence [9].

However, on a more positive note, there is also an opportunity to design places that can cope with the population growth by considering the social value and create improved outcomes for individuals as well as broader communities [10–18]. Within urban areas this can be in the form of more open green spaces and a diversity of green spaces within an area. This has been found to reduce stress levels and improve general fitness levels for people who use those spaces [19]. Even just having a view of green space has been shown to provide improved well-being benefits and create a range of other social values [17]. Within buildings, having shared communal spaces that are not crowded or noisy have been found to promote informal social connections and ties that can develop into friendships [9]. Previous research has highlighted five key areas within the built environment likely to promote well-being [20, 21]:

- having control over the internal environment, such as overheating, lighting, noise, dampness, and draughts;
- the quality of housing design and maintenance;
- having valued facilities, such as green spaces and social and community facilities;
- low crime rates and little fear of crime; and
- social participation, through having places and events to connect.

Population growth in cities has led to the increase of multiple occupancy buildings (such as high-rise apartments) being constructed. This form of higher-density living represents both a challenge and an opportunity to create buildings that prioritise social value. The amenities, sustainability, safety, and security of these buildings are amongst the social value considerations that can positively influence the living experience of residents. For example, the multiple-award-winning Granville New Homes regeneration project in London has been widely acknowledged for the quality of its design. Completed in 2009, the project engaged residents throughout the design and construction process – which has been highlighted as a key step in delivering social value in the built environment [22]. The Granville Development accommodates 110 dwellings, along with a youth centre, and public and resident-controlled communal outdoor spaces [23]. It also has landscaped communal gardens and streetscape to private patios, balconies, and roof terraces. The building has been acknowledged for its sustainability, security, and living spaces that are generously sized and provide good lighting throughout.

Another example is the Nightingale Housing model in Australia [10, 24]. This housing model started in Melbourne with the design and construction of their first building, The Commons, and the model has formally been used to inform the design and construction of at least 17 other developments, with other developments emerging that follow their model informally [25]. The Nightingale Housing model has five key guiding principles: affordability, transparency, sustainability, deliberative design, and community contribution. In effect it has flipped higher density living in Australia on its head and moved from a focus on economically efficient construction to placing the occupant and improving their well-being and social and economic value at the heart of what they do [25]. The removal of individual apartment laundries and having a shared communal laundry on the roof of the building is one example which reduces the costs of the project but importantly was done to drive those casual social interactions between residents in the building. As lead architect of the first few developments and pioneer of the Nightingale Housing Model Jeremy McLeod of Breath Architecture states, 'when you are doing your washing on the rooftop you quickly meet all your neighbours. Meeting people over washing laundry is a good way to break down barriers pretty fast. After that happens a few times, there are no awkward silences!' [10].

These two examples are of building projects that have attempted to go beyond basic requirements with socially sensitive designs to enhance the experience for people within the built environment. For people living in well-designed and constructed buildings, their house becomes a home. It is a place that has social value and where the safety, health, and well-being of the residents can prosper. We explore this further in Chapter 9.

To achieve high levels of social value for the people and communities within the built environment, the design and construction of structures need to go beyond 'fit for purpose' and 'minimum standards'. However, there has typically been resistance by the building industry to go beyond minimum expectations in construction project delivery [26–30]. The way the industry is structured means there is a motivation for companies and contractors to bid low to win work and as a result, quality, performance, and safety outcomes can be reduced, with construction projects just trying to make ends meet financially while meeting minimum standards and expectations [31]. The reluctance to invest and an intent to only meet minimum standards means that in some cases, the minimum standards are not even met, as projects do not go to plan. In other cases, completed construction projects may initially meet minimum requirements, but within months or years, defects begin to emerge, and there are again quality and safety issues to cope with post-construction.

Thus, construction projects should be considered beyond the initial completion of the construction phase, to ensure that, at the very least, minimum requirements and expectations are able to be met across an expected life of the building (or at least some subset of that). When human-made structures within the built environment do not meet basic requirements, it can be detrimental for end-user safety, health, and well-being. This is despite attempts to protect consumers of the built environment through housing policies and building warranties. One high-profile example of this is the Opal Tower building in Sydney, Australia (see Case Study 4.1).

Case Study 4.1   Evacuated for Christmas: Opal Tower (Australia).

On Christmas eve (24th December) of 2018, there was a mass evacuation of a 34-storey building containing 392 residential apartments in Sydney, Australia. Loud banging had been reported by residents, prompting fears of structural collapse [32]. At midnight, residents were informed they could return, only to be evacuated again on the 27th of December [33].

Over the following months, inspections, reports, and rework were undertaken to allow for residents to return intermittently in waves throughout 2019. However, there were some delays following disputes between the builder and developer over the remaining work to be completed [34]. Finally, after a turbulent year, all Opal Tower residents had returned in time for Christmas 2019 [35].

During many months of 2019, residents were forced into temporary accommodation with their lives feeling on hold. The implications were also felt by investors, who could not rent their property in the Tower, causing financial and mental stress. There were also concerns about the value of their properties, with owners wondering who would want to buy their apartment and live in Opal Tower. There would need to be confidence placed into potential buyers that the building was structurally sound and under robust warranty. These financial challenges and uncertainties can have a significant influence on the mental health of those that have invested capital into the defective properties.

While there has been a focus on the financial costs and who should absorb these costs, there is also more at stake for residents of the defective properties. Those in temporary accommodation may feel like they have their life on hold, as they cannot live in their home, sell their home, start a family in their home, and so forth. These challenges that affect the well-being of homeowners of defective properties are not considered during the rectification periods. It is clear in this Opal

Tower case that the real cost of defects will go beyond the financial uncertainty, as hundreds of residents had to be shifted from their homes and live 'in limbo'. The case also highlights that as we move increasingly to higher density living, any issues that do arise will impact a larger number of residents, further exacerbating issues relating to their well-being and the opportunities or solutions to address the issues.

4.2   Dealing with defects: Implications for well-being

Buyers of residential homes should be adequately protected through consumer rights. It is through this legal framework that post-construction housing quality is determined and outlined in an attempt to ensure that minimum standards are met. These consumer rights are often in the form of building warranties, which differ internationally in terms of juration and coverage. For instance, in New Zealand, the United Kingdom, and parts of Canada a ten-year warranty is provided [36–38]. However, in Australia the warranty period is between 6 and 12 years, depending on the local states and territories [39].

Typically, a building warranty will cover the new residential property for general defects and structural issues. Different governments may also outline pathways for addressing warranty claims; however, this can often be a long and difficult process for homeowners, who may have to invest significant time and financial input to have their claims heard. When dangerous defects that have daily implications for residents are ignored, delayed, or contested, the emotional toll on homeowners, who have invested life savings into properties, can be detrimental to their safety, health, and well-being. This has been reinforced by past studies, which highlighted that pursuing legal action to try and force rectification of defects is a long, stressful, and expensive experience [40]. Many owners' that choose this path experience adverse health consequences and do not recover the full financial costs, even if the legal pursuit is successful [41]. A recent analysis of defect dispute resolutions has revealed significant challenges for homeowners to resolve defect disputes [42]. The study highlighted the inadequacy of the current regulatory framework.

Our previous work has also reinforced these findings and also highlighted that homeowners are often dealing with multiple defects at the same time, which adds a range of complexity to the process. As one homeowner stated in a research interview:

I think a lot of the people are finding it [living in the apartments] really stressful, because you just don't know what is going on, you don't know how much you're going to have to pay, and there's always crisis after crisis to deal with … Our [water] tank is rusting away, that had to get replaced. There is ongoing maintenance disasters that costs us more money. We had a fire, and we had a flood about four months later … the sprinklers literally emptied the whole tank on the building and destroyed the lifts, because it just flooded straight out of the lift well. We had insurance for it, but some people didn't get to move back into their apartments for a good eight/nine months.

In other industries, such as the automotive, when your new car has a fault, you can return it to the place of purchase, and it is typically fixed quickly and effectively. Yet, in the construction industry, this type of prompt and efficient customer service is few and far between. To become a more consumer-facing industry and enhance social value, the emphasis on ensuring building quality should not only be in the initial construction phases but also at any future point while the building remains under warranty. Thus, future policy and practice solutions should also consider the customer service during any building warranty claim, as well as clarity on where in the construction supply chain accountability for the defect lies and where the costs for rectification should be attributed. The customer service throughout the period from the initial emergence of the defect to rectification needs to be further considered in policy and practice, to shift towards a more consumer-facing construction sector.

Defects can occur in the construction phase and remain undetected until the building is occupied. Alternatively, the defect can emerge weeks, months, or years after building completion. Dealing with such defects can be particularly challenging, with many developers and builders trying to avoid having to return to rectify the issues. At this point, disputes can arise, and it is not clear who will absorb the financial costs.

The financial costs have been almost the sole focus in solving the problems that arise from housing defects. However, previous research suggests that there should be broader considerations to provide a holistic and comprehensive response to residential defects [41, 43, 44]. For those experiencing defects in their home, there can be a significant reduction in their liveability and well-being, due to challenging experiences of residing in their home. For example, one interviewee in our previous research, stated:

> What more defects are we going to have to deal with? And I am feeling very burnt out from trying to engage with them [builders] over all of this stuff – I've had enough. This has been going on, not just for a couple of months. This has been going on year in and year out… I am feeling very threatened, very intimidated. I am really feeling very unsafe in my home. I'm feeling very distressed and very depressed about things.

The challenges that emerge from building defects and disputes can be emotionally exhausting and very stressful. The journey to defect rectification can also be long, complex, and often with an unclear endpoint. There is far more to be considered here than simply the financial cost to fix and replace defects, which has been demonstrated throughout the many building crises that have occurred across the world. For example, those that lived and are living through the leaky homes' saga in New Zealand (see case study in Chapter 2). This saga has been labelled as New Zealand's largest human-made disaster, which emerged around the 1990s and there are still owners almost walking away from their leaking units, with some apartments in Auckland selling for as little as NZ$20,000 [45]. The 30 years of defects, danger, disputes, and distress is still not over, with scepticism still over new builds in New Zealand.

Similar leaky buildings have been reported elsewhere in the world, with, for example, the leaky condo crisis in Canada, which is also still an ongoing issue despite problems arising in the late 1980s (see case study in Chapter 8). With substantial water ingress, there are safety concerns, as the timber frames will slowly rot; and there are health concerns, as mould will grow in the dark and damp conditions around the rotting timber frames. As highlighted repeatedly in this chapter, there are also mental health and well-being concerns, as consumers must live with the everyday stresses associated with being unable to rectify such defects in a timely manner. The latest dangerous building defect to affect consumer safety, health, and well-being is in the form of combustible cladding.

4.3 A contemporary dangerous defect: Combustible cladding

In 2017, the Grenfell Tower disaster in the United Kingdom (see case studies in Chapters 5 and 7) brought scrutiny to combustible cladding and the construction sector. There were over 70 fatalities and 70 injuries in the Grenfell disaster, where combustible cladding helped fuel a large fire up the façade of the tower very quickly [46]. Cladding is when one layer of material is applied over another, to provide another layer or 'skin'. It can provide thermal, acoustic, and aesthetic benefits for the building. However, in some cases, the cladding applied to buildings has been found to be dangerously flammable.

Combustible cladding (such as the form on Grenfell Tower) is not only a threat to residents' safety, but also a threat financially. It is a costly defect to replace or rectify and the rectification process can often reveal a range of other defects that also need to be fixed. Homeowners have found that they may have to absorb most, or all, of the costs (see Chapter 3). This can have significant implications for mental health and well-being, as there could be stress that manifests from these safety and/or financial concerns. Our research in Australia [47] revealed that homeowners typically had no idea that they had flammable cladding on their buildings or that they would be responsible for addressing this defect. For example, one homeowner stated:

> We had never even heard of flammable cladding 10 years ago when we first moved in.

The initial shock and surprise for many that they had flammable cladding would soon change to negative emotions as homeowners struggled to rectify the defect. For those that were in low-risk buildings, or were financially secure, these emotions were more likely to be frustration and annoyance. However, for others that were not as financially secure or were concerned for their safety, they were anxious, stressed, and even distraught. For instance, one homeowner explained how others in her building were failing to cope and expressed significant concerns for their well-being:

> There's certainly one lady, who's even worse off than me, and she's absolutely beside herself. She doesn't know what's she's going to do … She'll probably go insane or kill herself or something. Seriously, and I'm saying that seriously. She is so distraught by this all.

There were also feelings of anger and injustice, as homeowners believed it was not their fault they were in this situation, and yet they were the ones being affected by the combustible cladding. Many had followed recommended processes when purchasing their properties such as having a building inspection undertaken, yet those processes had not revealed this issue. As one homeowner put it:

> This is a human issue. I'm heartbroken.… How could they do this to me?.

The emotions homeowners displayed varied in type and magnitude, but they were all negative. Having positive emotions is one of the more obvious components of enhanced well-being, and on the other side of the scale, negative emotions reduce well-being. Thus, the negative feelings and emotions that had manifested from the combustible cladding situation were having implications for homeowner well-being.

It is important to note that well-being is not only about the absence of negatives, including depression or anxiety, but also positives that can promote a healthy and happy life. Homeowners explained that regarding many of the activities that brought them joy, such as forms of entertainment, dining out, and going on holiday, they need to reconsider due to the financial uncertainty of the cladding rectification costs. For example, one owner stated:

> You have to think about costs because of the uncertainty … So there will be cut-backs…Certainly it will cut-back on other discretionary spending that we could have spent, either on eating out, or going on holidays.

For different homeowners this had varying implications depending on their financial situation. Those more financially stable may have chosen to avoid further investments to reduce financial risks and ride out any potential costs from their apartments with flammable cladding, whereas those that were less financially secure were making decisions to strengthen their financial situation such as delaying retirement, not going out for entertainment, or opting for a short local holiday, rather than a long international trip. Clearly, not being able to do the activities that people enjoy as often as they normally would has implications for their well-being. Having to constantly consider the potential financial implications of combustible cladding was emotionally exhausting and had impacts on well-being outcomes.

Another important component of well-being is having good relationships with others. Various types of relationships can help promote well-being, from a temporary friendship that lasts briefly on a holiday to longer-term relationships with friends, family, and partners. As highlighted at the beginning of this chapter, the built environment can play a role in creating spaces that encourage social interactions where these relationships can begin, grow, and develop. In buildings with combustible cladding, many of these communal areas where social interactions would normally take place were closed to lower fire safety risks. This closure was because at gatherings there was often an ignition source (e.g. barbeques, smoking) and a high density of people that potentially had a long journey to escape the building (e.g. rooftop gathering) should a fire start. The loss of a place to congregate within multiple occupancy buildings clearly reduced the social value. It was described as a real inconvenience by residents, who enjoyed spending time in these communal areas. For instance, one homeowner stated:

> There is a roof garden with barbecues and that's been closed for nearly a year because of the cladding issue. So that is a real inconvenience ... A lot of us don't get a lot of sun and people go up there and meet and have parties and we've got a garden club ... We have a garden and only one person is allowed to go, under the rules that they've established.... Only one person from the garden club goes up to mind the garden. She puts pictures on Facebook.

The issue of cladding also negatively affected those that were trying to solve the problem and organise rectification of the defect. Typically, these were homeowners that were also on their owners corporations, and they were tasked with decisions and actions to step towards cladding rectification (see Chapter 6). Being involved in voluntary organisations can enhance well-being, as it is way for people to have purpose, meaning, a sense of engagement, and accomplishment. However, negative implications for well-being are also possible when there is an absence of these. This was the case with homeowners of combustible cladding within the owners corporations. Since there was a lack of accomplishment, with buildings not being rectified and no clear actions to take steps forward, creating a lack of purpose, meaning, and engagement for those with the owners corporations. For example, one homeowner within our research study stated:

> There is no particularly obvious way forward that seems to be the definite cause of action to take ... because deeming a building to be safe from cladding, we don't know if that means taking all the cladding off, or if that means making sure that the fire exits have no cladding near it, or if it means something else.

The mental stress that accompanied these challenges with rectification is also significant, as homeowners expressed concerns about potential financial costs and safety risks. The magnitude of these implications can be understood from the situation in the United Kingdom, where more and more buildings are being identified with combustible cladding, which is seriously affecting the health and well-being of many homeowners and tenants. The challenges with rectification have seen those affected unite and form various cladding action groups, such as the United Kingdom Cladding Action Group. The case study example below (Case Study 4.2) highlights some findings reported by the United Kingdom Cladding Action Group:

Case Study 4.2   The combustible cladding crisis in the United Kingdom.

The United Kingdom Cladding Group Action Group surveyed 550 homeowners with combustible cladding on their buildings [48]. The survey found that as a direct result of the situation in their building:

- Nine out of ten had deteriorating mental health.
- Eight out of ten had strained relationships with friends and families.
- Seven out of ten reported difficulties sleeping.
- 46% were planning to medical help.
- 28% had put starting a family on hold.
- 23% reported suicidal feelings or a desire to self-harm.

The following quotes from Twitter users retweeted by Cladding Action Groups provide insight into the distressing situations many homeowners face. For example, the following tweets were made in 2020 (from three different homeowners):

My entire life is on hold because a flat I bought has cladding on it. Can't move, can't change jobs, can't get married, can't have a baby, can't spend because I will be made bankrupt otherwise.

Lost two stones in weight, cannot sleep, worried sick about how I'm going to find money [for cladding bills].

My son. He is 2. Trapped in one bed flat. Our lounge is: His bedroom. His play area. our dining room. We can't buy toys: no space. No £for his future. Tell him why we are trapped here. What did mummy and daddy do wrong to end up like this?

4.4 Next steps: A shift in policy focus?

Broadly speaking, policy has traditionally been based on principles and perspectives that attempt to improve and strengthen countries and communities economically. However, over the last couple of decades there has been acknowledgement for world leaders that economics should not be the only focus. For example, former British Prime Minister David Cameron stated:

Too often in politics today, we behave as if the only thing that matters is the insider stuff that we politicians love to argue about – economic growth, budget deficits and GDP. Gross domestic product. Yes, it's vital. It measures the wealth of our society. But it hardly tells the whole story. Wealth is about so much more than pounds, or euros or dollars can ever measure. It's time we admitted that there's more to life than money, and it's time we focused not just on GDP, but on GWB – general well-being.

(David Cameron [2006], British Prime Minister [2010–2016])

Government policy from an economic perspective has previously assumed that well-being will be enhanced as a direct consequence of a stronger economy. However, the inverse can also be argued: that enhancing well-being will result in a stronger economy. This is based on the evidence that billions of pounds, euros, and dollars are lost every year due to mental health issues, such as stress, depression, and anxiety. For example, it has been estimated that in Australia, mental health issues cost businesses almost AU$11 billion every year in the form of absenteeism, presenteeism, and compensation claims [49]. In Canada, lost productivity, health care, and reductions in health-related quality of life cost the country an estimated $51 billion every year [50, 51]. In the United Kingdom, mental health is the single largest cause of disability and costs an estimated £105 billion in the form of lost workplace productivity, service costs, and reduced quality of life [52]. Thus, a greater focus on well-being in policy development could not only improve well-being but also reduce these financial costs and improve countries economically too. It is therefore worth considering a more central focus on well-being in policy development, which represents a new way of thinking and, arguably, a better approach [53].

This call for greater integration of health and well-being has been debated in wider policy development discussion [54], as well as within the building and housing space [55]. Within the built environment, policy has aligned with broader government approaches through a focus on policy from an economic perspective. While rethinking and creating policy that has well-being as a priority is challenging, the built environment plays a very important role in enhancing well-being and therefore it can be contended that this line of thought has a sound basis. An example of how policy within the built environment can improve health and well-being is through policies that encourage the design and construction of green spaces or that reduce building on green spaces. These policies, which create and keep green areas close to and within cities, provide a range of health and well-being benefits including:

- encouraging exercise [56],
- fostering greater social cohesion in neighbourhoods [57],
- less risk of psychological distress [58],
- improved air quality [59], and
- reduced risks of morbidity [60].

This is one example where the built environment can significantly enhance health and well-being. Thus, governments should consider having a more widespread policy that encourages the design and construction of healthy cities.

There should also be policy that protects the well-being of people when issues and defects emerge. The World Health Organisation highlighted that for healthy urban areas, there needs to be a strategic approach to manage low-quality housing [61]. Within the WHO European region there are over 100,000 deaths per year due to inadequate housing conditions [61]. The majority, if not all, of these poor conditions are preventable. In regard to mental health, it is widely acknowledged that poor-quality housing appears to increase psychological distress, but research methodology challenges make it difficult to draw clear and definitive conclusions [62]. Poor housing conditions can emerge when defects arise within the built environment. There is then a risk that those living within the buildings will be affected, especially when the defect is left unresolved and can threaten the health or safety of residents. There are many high-profile examples of such cases, with combustible cladding being the latest defect to significantly affect the health and well-being of residents.

Considering how defects damage the health and well-being of residents, there must be more careful consideration into housing policy that protects homeowners – not only financial but also mentally. Currently, housing policy is developed from an economic perspective, where building warranties attempt to protect homeowners financially by outlining that defects under warranty should be fixed at the expense of those responsible. As is demonstrated from the combustible cladding case examples, this financial protection does not always work as intended and often leaves homeowners to address these issues on their own (see also Chapters 3 and 6). However, even in cases where the financial protection through building warranties acts as it should, there is still no consideration for the emotional toll homeowners must go through to reach rectification. This process can be a long and emotionally draining experience that has serious implications for homeowner health and well-being. Thus, housing policy should not only focus on tightening up issues with financial protection but must also consider well-being to provide more comprehensive consumer cover. Hence, the effectiveness of housing policy and building warranty should be understood through not only whether the defects are rectified or not but also the time taken to resolve the issues and the levels of consumer satisfaction throughout this process.

This would go beyond the focus on economic-based housing policy, since it would also capture the emotional experience consumers go through during the rectification process. Policy that does consider well-being would provide a more comprehensive response than current approaches that largely focus on the economics of defects. This policy reform would then acknowledge that the real cost is much more than the potential financial losses but also costs to livelihoods, social value, and overall well-being of residents.

4.5 Conclusions

The built environment can enhance well-being by considering social value through the design and construction of buildings and cities. However, when issues around design, quality, performance, and defects arise, then the built environment also has the potential to reduce well-being. This can particularly be the case with homeowners, as they invest considerable financial capital and spend considerable time within their dwellings.

The financial cost of high-profile defects, such as combustible cladding, is not the only cost the homeowners are exposed to. The long and emotional journey from the emergence of the defect to rectification can significantly affect the well-being of homeowners. Fears of asbestos, structural collapse, and cladding fires have all occurred in residents with dangerous home defects. While the financial cost of the rectification work can be calculated, the human cost to those that are stressed, anxious, live in fear, or must evacuate, has not received the attention it deserves.

A more comprehensive and holistic approach to managing defects is required that goes beyond policy that focuses only on the economic costs. While having building warranties and financial protection for homeowners is an important policy, there should be greater consideration into how to create processes, policies, and frameworks that encourage a more consumer-facing construction sector, for example, having a process in place that allowed for a clear and efficient way to rectify buildings within warranty would significantly reduce mental stresses placed upon homeowners. Further discussion on such ideas to improve government response and policy is in Chapters 8 and 9.

To improve the response to high-profile defects, it is recommended that further research and work is undertaken to consider creating a more comprehensive policy response. This response should not only focus on the economics of fixing and replacing defects but also include the customer satisfaction throughout this process. A more punctual, functional, and efficient customer-facing response to dangerous defects would lead to enhanced health and well-being outcomes. Otherwise, homeowners will continue to be left with mental stress and other potential health implications that manifest from defects for long and uncertain periods.

While government policy with a greater consumer focus would represent an important step forward, there is also a reliance on industry to be part of the solution and operate with greater corporate social responsibility – which is discussed in the following chapter.

References

1. Moore, T., F. de Haan, R. Horne, and B. Gleeson, *Urban sustainability transitions. Australian Cases – International perspectives.* 2018, Singapore: Springer.
2. Klepeis, N.E., W.C. Nelson, W.R. Ott, J.P. Robinson, A.M. Tsang, P. Switzer, J.V. Behar, S.C. Hern, and W.H. Engelmann, *The national human activity pattern survey (NHAPS): A Resource for assessing exposure to environmental pollutants.* Journal of Exposure Analysis and Environmental Epidemiology, 2001. **11**(3): pp. 231–252.
3. Moore, T., L. Nicholls, Y. Strengers, C. Maller, and R. Horne, *Benefits and challenges of energy efficient social housing.* Energy Procedia, 2017. **121**: pp. 300–307.
4. CABE, *The value of good design. How buildings and spaces create economic and social value.* 2002, London: CABE.
5. Sherriff, G., P. Martin, and B. Roberts, *Erneley close passive house retrofit: Resident experiences and building performance in retrofit to passive house standard.* 2018, University of Salford: Salford.

6. Kats, G., *Greening our built world: Costs, benefits, and strategies.* 2009, Washington, DC: Island Press.
7. Willand, N., C. Maller, and I. Ridley, *Addressing health and equity in residential low carbon transitions – Insights from a pragmatic retrofit evaluation in Australia.* Energy Research & Social Science, 2019. **53**: pp. 68–84. DOI: 10.1016/j.erss.2019.02.017.
8. World Health Organization, *WHO European Healthy Cities Network.* 2020, WHO, 12/23/2020 https://www.euro.who.int/en/health-topics/environment-and-health/urban-health/who-european-healthy-cities-network#:~:text=WHO%20Healthy%20Cities%20is%20a,and%20approximately%2030%20national%20networks
9. Sullivan, W., and C. Chang, *Mental health and the built environment.* In: *Making healthy cities*, 2011. Island Press: Washington. pp. 107–117.
10. Moore, T., and A. Doyon, *The uncommon nightingale: Sustainable housing innovation in Australia.* Sustainability, 2018. **10**(10): p. 3469.
11. De Laurentis, C., M. Eames, and H. Hunt, *Retrofitting the built environment 'to save' energy: Arbed, the emergence of a distinctive sustainability transition pathway in Wales.* Environment and Planning C: Government and Policy, 2017. DOI: 10.1177/0263774X16648332.
12. Hagbert, P., and K. Bradley, *Transitions on the home front: A story of sustainable living beyond eco-efficiency.* Energy Research & Social Science, 2017. **31**: pp. 240–248. DOI: 10.1016/j.erss.2017.05.002.
13. Boyer, R., *Grassroots innovation for urban sustainability: Comparing the diffusion pathways of three ecovillage projects.* Environment and Planning A: Economy and Space, 2015. **47**(2): pp. 320–337. DOI: 10.1068/a140250p.
14. Ridley, I., J. Bere, A. Clarke, Y. Schwartz, and A. Farr, *The side by side in use monitored performance of two passive and low carbon Welsh houses.* Energy and Buildings, 2014. **82**: pp. 13–26. DOI: 10.1016/j.enbuild.2014.06.038.
15. Newton, P., *Liveable and sustainable? Socio-technical challenges for twenty-first-century cities.* Journal of Urban Technology, 2012. **19**(1): pp. 81–102. DOI: 10.1080/10630732.2012.626703.
16. Yudelson, J., *The green building revolution.* 2010, Island Press.
17. CABE, *The value of good design. How buildings and spaces create economic and social value.* 2002, London: Commission for Architecture and the Built Environment. https://www.designcouncil.org.uk/sites/default/files/asset/document/the-value-of-good-design.pdf
18. Montgomery, C., *Happy city: Transforming our lives through urban design.* 2013, UK: Penguin.
19. Grahn, P., and U. Stigsdotter, *Landscape planning and stress.* Urban Forestry & Urban Greening, 2003. **2**(1): pp. 1–18.
20. Chu, A., A. Thorne, and H. Guite, *The impact on mental well-being of the urban and physical environment: An assessment of the evidence.* Journal of Mental Health Promotion, 2004. **3**(2): pp. 17–32.
21. Guite, H., C. Clark, and G. Ackrill, *The impact of the physical and urban environment on mental well-being.* Public Health, 2006. **120**(12): pp. 1117–1126.
22. De Sousa, S., *Building social value into design and placemaking.* In: *social value in construction.* 2019, Oxon: Routledge: Taylor and Francis. pp. 147–165.
23. Bernstein, L., *Granville new homes, Brent.* 2020, 10/12/2020 http://www.levittbernstein.co.uk/portfolio/granville-new-homes/
24. Doyon, A., and T. Moore, *The acceleration of an unprotected niche: The case of nightingale housing, Australia.* Cities, 2019. **92**: pp. 18–26. DOI: 10.1016/j.cities.2019.03.011.
25. Nightingale Housing. *What is the nightingale model.* 2018, 04/17/2018 http://nightingalehousing.org/model/
26. Moore, T., S. Berry, and M. Ambrose, *Aiming for mediocrity: The case of Australian housing thermal performance.* Energy Policy, 2019. **132**: pp. 602–610. DOI: 10.1016/j.enpol.2019.06.017.

27. Berry, S., T. Moore, and M. Ambrose, *Flexibility versus certainty: The experience of mandating a building sustainability index to deliver thermally comfortable homes.* Energy Policy, 2019. **133**: p. 110926. DOI: 10.1016/j.enpol.2019.110926.
28. Evans, M., V. Roshchanka, and P. Graham, *An international survey of building energy codes and their implementation.* Journal of Cleaner Production, 2017. **158**: pp. 382–389. DOI: 10.1016/j.jclepro.2017.01.007.
29. International Energy Agency, *Modernising building energy codes to secure our global energy future.* 2013, Paris: International Energy Agency. http://apo.org.au/node/192361
30. Lowe, R., and T. Oreszczyn, *Regulatory standards and barriers to improved performance for housing.* Energy Policy, 2008. **36**(12): pp. 4475–4481. DOI: 10.1016/j.enpol.2008.09.024.
31. Oswald, D., D. Ahiaga-Dagbui, F. Sherratt, and S. Smith, *An industry structured for unsafety? An exploration of the cost-safety conundrum in construction project delivery.* Safety Science, 2020. **122**: 104535.
32. Unisearch, *Opal tower investigation – Final report: Independent advice to NSW minister for planning and housing.* 2019, Sydney: NSW Government.
33. Lehmann, M., *Inside Opal Tower debacle.* 2019, Australia: The Australian, 12/11/2020 https://www.theaustralian.com.au/subscribe/news/1/?sourceCode=TAWEB_WRE170_a&dest=https%3A%2F%2Fwww.theaustralian.com.au%2Fweekend-australian-magazine%2Ffalling-through-the-cracks-sydneys-opal-tower-debacle%2Fnews-story%2Fb6f780d50823388feafe43f099e63702&
34. Dye, J., *Remaining Opal Tower residents to return home by Christmas.* 2019, Sydney Morning Herald, 12/11/2020 https://www.smh.com.au/national/nsw/remaining-opal-tower-residents-to-return-home-by-christmas-20190920-p52ser.html
35. Sas, N., *Opal Tower residents home for Christmas but fight for compensation looms.* 2019, ABC News, 12/11/2020 https://www.abc.net.au/news/2019-12-19/opal-tower-residents-back-home-but-conflict-remains/11811468
36. Ministry of Business, Innovation and Employment, *Issues after your building work has finished.* 2020, Ministry of Business, Innovation and Employment, 11/26/2020 https://www.consumerprotection.govt.nz/help-product-service/home-renovation-repair/issues-after-building-work-finished/#:~:text=10%2Dyear%20implied%20warranty%20period,that%20they%20don't%20apply
37. Home Owners Alliance, *New Home Warranties – What they do and don't cover.* 2020, 11/26/2020 https://hoa.org.uk/advice/guides-for-homeowners/i-am-buying/new-home-warranties-cover/
38. Alberta, G., *New home warranty.* 2020, 11/26/2020 https://www.alberta.ca/new-home-warranty-overview.aspx
39. Victoria, C.A., *Implied warranties and domestic building insurance – checklist.* 2020, 11/26/2020 https://www.consumer.vic.gov.au/housing/building-and-renovating/checklists/implied-warranties-and-domestic-building-insurance
40. City Futures Research Centre, *Regulation of building standards, building quality and building disputes.* 2019, Sydney: University of New South Wales.
41. Anonymous, *Strata defects, case study: Anonymous.* n.d., University of New South Wales: Sydney. https://cityfutures.be.unsw.edu.au/documents/484/Case_Study_Poster_Defects_4.pdf
42. Paton-Cole, V., and A. Aibinu, *Construction defects and disputes in low-rise residential buildings.* Journal of Legal Affairs and Dispute Resolution in Engineering and Construction, 2021. **13**(1): 05020016.
43. Oswald, D., T. Moore, and S. Lockrey, *Flammable cladding and the effects on homeowner well-being.* 2021, Housing Studies.
44. Oswald, D., T. Moore, and S. Lockrey, *Combustible costs! financial implications of flammable cladding for homeowners.* International Journal of Housing Policy, 2021.

45. Dyer, P., *Our leaky buildings saga is a long way from sorted.* 2019, Radio New Zealand, 12/15/2020 https://www.rnz.co.nz/programmes/the-detail/story/2018723459/our-leaky-buildings-saga-is-a-long-way-from-sorted
46. BBC News, *Grenfell Tower: What happened.* 2019, British Broadcasting Corporation, 12/14/2020 https://www.bbc.com/news/uk-40301289
47. Oswald, D., *Homeowner vulnerability in residential buildings with flammable cladding.* Safety Science, 2021. **136**. DOI: 10.1016/j.ssci.2021.105185.
48. The Leaseholders Charity, *Mental health report.* 2020, London: The Leaseholders Charity.
49. PwC, *Creating a mentally healthy workplace: Return on investment.* 2014, Melbourne: PricewaterhouseCoopers.
50. Smetanin, P., D. Stiff, C. Briante, C.E. Adair, S. Ahmad, and M.T. Khan, *The life and economic impact of major mental illnesses in Canada: 2011 to 2041.* 2011, Toronto: RiskAnalytica, on behalf of the Mental Health Commission of Canada.
51. Lim, K., P. Jacobs, A. Ohinmaa, D. Schopflocher, and C. Dewa, *A new population-based measure of the economic burden of mental illness in Canada.* Chronic Diseases in Canada, 2008. **28**(3): pp. 92–98.
52. Department of Health, *No health without mental health: A cross-government mental health outcomes strategy for people of all ages.* 2011, London: Department of Health.
53. New Economics Foundation, *Wellbeing in four policy areas.* 2014, London: New Economics Foundation.
54. Oishi, S., and E. Diener, *Can and should happiness be a policy goal?* Policy Insights from the Behavioral and Brain Sciences, 2014. **1**(1): pp. 195–203.
55. Clapham, D., *Happiness, well-being and housing policy.* Policy & Politics, 2010. **38**(2): pp. 253–267.
56. Giles-Corti, B., M.H. Broomhall, M. Knuiman, C. Collins, K. Douglas, K. Ng, A. Lange, and R.J. Donovan, *Increasing walking: How important is distance to, attractiveness, and size of public open space?* American Journal of Preventive Medicine, 2005. **28**(2 supplement 2): pp. 169–176.
57. Kaźmierczak, A., *The contribution of local parks to neighbourhood social ties.* Landscape and Urban Planning, 2013. **109**(1): pp. 31–44.
58. Francis, J., L.J. Wood, M. Knuiman, and B. Giles-Corti, *Quality or quantity? Exploring the relationship between public open space attributes and mental health in Perth, Western Australia.* Social Science & Medicine, 2012. **74**(10): pp. 1570–1577.
59. Nowak, D.J., S. Hirabayashi, A. Bodine, and E. Greenfield, *Tree and forest effects on air quality and human health in the United States.* Environmental Pollution, 2014. **193**(October): pp. 119–129.
60. Maas, J., R.A. Verheij, S. de Vries, P. Spreeuwenberg, F.G. Schellevis, and P. Groenewegen, *Morbidity is related to a green living environment.* Journal of Epidemiology & Community Health, 2009. **63**: pp. 967–973.
61. World Health Organization. *Cities: Urban planning and health, fact sheet 2.* In: *Sixth Ministerial Conference on Environment and Health.* 2017, Ostrava: WHO.
62. Evans, G., *The built environment and mental health.* Journal of Urban Health, 2003. **80**(4): pp. 536–555.

5   Corporate social responsibility for the consumer

Key takeaways

- Corporate social responsibility (CSR) is about companies within the built environment being socially accountable to a range of relevant stakeholders, including the housing consumer.
- There are several examples of where basic CSR requirements are not met, such as the case of combustible cladding, where there has been a lack of response and support provided to homeowners and limited rectification work completed to date.
- Governments need to take a more active role in ensuring robust policies that protect the housing consumer, as well as having improved enforcement of building codes and legislations to avoid egregious outcomes, such as the Grenfell disaster.
- Companies seeking to provide higher levels of CSR need to carefully consider a holistic strategy that involves every stakeholder across the life cycle of their projects.

Chapter summary

This chapter discusses the concept of corporate social responsibility (CSR) in the built environment, with a particular focus on the consumer. Construction companies that create the built world have often overlooked the housing consumer's needs and expectations, especially when these needs and expectations are conflicting with time and money. In the opening section of this chapter, the concept of CSR is discussed by reflecting on how this concept has developed over the last 50 years. Thereafter, the CSR challenges in the built environment are highlighted, while basic expectations including for the consumer are explicitly clarified. The devastating impacts of disregard for CSR are highlighted through case examples, such as the Grenfell disaster. Finally, the later part of the chapter focuses on recent government action in Australia that has been undertaken to help protect consumers, where basic CSR expectations have repeatedly not been met.

It is recommended that further research is undertaken to assess the effectiveness of new government policy that is being implemented in an attempt to address

DOI: 10.1201/9781003176336-5

consumer protection issues within the built environment. Further, the companies within the built environment should not just rely on what the government directs them to do as a minimum but seek to take greater social accountability for all stakeholders, especially the consumer. There are recommendations throughout this chapter of what stakeholders should be considered, and what this would look like, in terms of CSR initiatives and actions.

5.1 An introduction to corporate social responsibility

The concept of CSR originated around 50 years ago and has been undertaking somewhat of a revival with increasing attention in recent years. Over the years, CSR has stimulated debate around the role of corporate stakeholders and how they engage with wider society. It is also a concept that has been defined in various ways and has had a wide range of differing perspectives on how it can be achieved. This is because CSR is a nebulous and multidimensional concept, where it means different things to different organisations, making it difficult to universally define [1]. In short, CSR can be understood as a business model, where companies self-regulate to be socially accountable to their own employees, other stakeholders, and the public. Examples that are relevant across different industries include: participating in fair trade, charitable giving, and reducing carbon footprints. Within the built environment, CSR should be considered by different organisations, such as builders, architects, and engineers, in both the construction phases *and* post-construction. These considerations are about acting in the best interests of all stakeholders, including employees, consumers, and the public, rather than pursuing what is advantageous only for shareholders or senior management. It is very much about social considerations in any decision-making processes.

The argument for the use of CSR has been challenged over time, particularly when CSR was an emerging concept. In the 1960s, it had been claimed that the social responsibility of firms was to maximise shareholders profit through increasing wealth outcomes. In doing so, other social benefits, such as employment opportunities and workplace pensions, would manifest into the community. Thriving business was therefore viewed by some as achieving CSR. This meant other actions that might create social benefit were often dismissed, especially if they reduced financial profits. As Nobel Prize winner Milton Friedman stated [2]:

> there is one and only one social responsibility of business–to use it resources and engage in activities designed to increase its profits so long as it stays within the rules of the game, which is to say, engages in open and free competition without deception or frau.

Around 50 years on from this statement, it is now widely acknowledged that businesses should do more than simply maximise profits and meet minimum legal, industry, or financial requirements (e.g. do not commit deceit or fraud). Thus, the concept of CSR has now evolved to consider more than simply following minimum government regulations and the absence of poor ethical judgements.

Examples of poor ethical principles within the construction industry could be providing fraudulent advice or charging extortionate prices to consumers who do not have a good understanding of the work that is to be carried out. Companies that avoid these types of actions demonstrate some form of CSR by conforming to the basic expectations of society. However, to achieve high levels of CSR, a company should go above and beyond these expectations and be looking to lead the industry, and the wider community, to deliver improved social outcomes. To achieve this, organisations should have the mindset that they both have an economic contract and a social contract with society, where social goals should be closely considered, as well as economic targets. Some researchers and policy makers argue that the social contract should supersede any financial contract, such as the B Corp movement which has emerged in recent years [3–5]. That is not to say that companies should run at a financial loss but that social and ethical considerations should drive financial decisions rather than the other way around.

In the 1990s and 2000s, there was an inquest into the interconnection between social goals and economic goals in search of a business case for CSR [6]. By providing evidence of financial return for companies that demonstrated high levels of CSR, there was hope of a wider spread adoption of CSR business models. This is based on the realisation that while many companies have concern for social issues and 'doing the right thing', they are also driven by 'the bottom line' financially, particularly in the construction industry [7]. This is not unexpected given the neoclassical market, where financial growth is seen as the measure of a successful and strong company.

Research to prove the link that higher levels of CSR also have short- or long-term economic benefits, has not yet been clearly proven at the time of writing [6]. Therefore, further research is still required to improve understanding of the business case for CSR [8]. While this link that higher CSR produces higher profits is still unclear, there is an inverse connection: that companies that have higher profits tend to invest more into CSR initiatives. Thus, greater profit often results in higher levels of CSR; as opposed to higher levels of CSR resulting in greater profits. This indicates that companies that do make a significant profit are willing to invest in CSR opportunities, likely because they have the capital available to do so. Unfortunately, the way the construction industry is structured means that low profit margins, low bidding, and cost-cutting to stay competitive are all too common. This means that there is less likely to be such investment in CSR initiatives or opportunities within the built environment.

5.2 CSR in the built environment

There has been increasing pressure for those constructing the built world to act as good corporate citizens that are socially responsible. This has come as part of a wider push for the built environment to deliver improved affordability, quality, performance, and environmental outcomes. Research into CSR in the construction industry is growing [6], as is the importance for construction companies to have a good CSR record [9]. However, construction companies are still in an emerging and immature CSR state, where work practices have been widely criticised and are disconnected from stated ambitions for CSR objectives. In particular, there has been an increasing focus on hidden global issues, such as modern slavery, which have challenged the construction industry to do better [10]. For example, the Chartered Institute of Building (CIOB) released a report on 'the dark side of construction' [11], where the chief executive, Chris Blythe, stated:

> Our sector is rife with human rights abuses. Bonded labour, delayed wages, abysmal working and living conditions, withholding of passports and limitations of movement are all forms of modern slavery. And our business models must take a large part of the blame: the global trend towards outsourcing and cut price contracting makes it easy for main contractors to duck out of their responsibilities. The plight of the most vulnerable gets lost among the long and complex supply chains. It's too convenient to blame the subcontractor or poor local legislation. You might think that modern slavery is not a problem where you work. Think again. Human exploitation is a global issue, embedded both in the developed and developing world.

Critics of the construction sector have also extended concerns to other issues including poor consumer satisfaction, defects, cost and time overruns, and a lack of innovation and research for improvements [12–15]. Hence, there is a need for those operating within the built environment to consider CSR more closely, carefully, and holistically through a more genuine lens. The reluctance of some companies to focus attention on CSR is because there are some notable barriers to the greater uptake of CSR initiatives. These barriers can begin with a lack of clarity or dispute on how CSR is defined. Without a clear definition, it is not obvious what is within the scope of CSR and what objectives and goals should be strived for. There are also deeply entrenched ways of working that are challenging to change and there can be resistance from the supply chain, and scepticism on the return for investment due to the narrow focus on financial outcomes. While there have been some notable attempts to address these issues (either due to engaging with CSR or other drivers), there remains a lack of engagement from the industry as a whole. For example, the construction industry is notorious for not sharing intellectual property, which means every time a company wants to innovate or do something for the first time there is a lack of information from which they can draw upon. Furthermore, a lack of research understanding is also a barrier in outlining evidence-based approaches, as well as a lack of widespread leadership and management skills in this developing concept.

Despite these challenges there are various potential benefits for organisations that demonstrate high levels of CSR. These can consist of an improved reputation, broader market access, and higher rates of staff retention. Relations with stakeholders, such as the community, government, and suppliers, are also likely to develop and ideally lead to repeat business. Other reported benefits are greater risk management, innovation, abilities to respond to challenges, and overall competitiveness in the market.

The emergence of Certified B Corporations (or B Corps) is an example of businesses exercising CSR. B Corps are businesses that have demonstrated and meet 'the highest standards of verified social and environmental performance, public transparency, and legal accountability to balance profit and purpose' [16]. In meeting this higher verifiable standard B Corps are attempting to drive a global culture and business shift which is more inclusive, equitable, and sustainable and is not only better in terms of the products and services provided but all stakeholders involved and there are increasing examples of stakeholders in the built environment who are becoming certified B Corp [17]. The B Corp declaration of interdependence states [16]:

> We envision a global economy that uses business as a force for good. This economy is comprised of a new type of corporation – the B Corporation – Which is purpose-driven and creates benefit for all stakeholders, not just shareholders. As B Corporations and leaders of this emerging economy, we believe:
>
> - *That we must be the change we seek in the world.*
> - *That all business ought to be conducted as if people and place mattered.*
> - *That, through their products, practices, and profits, businesses should aspire to do no harm and benefit all.*
> - *To do so requires that we act with the understanding that we are each dependent upon another and thus responsible for each other and future generations.*

Within the built environment, there are certified companies listed as B Corps, including more than 60 in the architecture/design/planning category, 6 in the building materials' category, 28 in the contractors' and builders' category, and 16 in the design/build category. This list includes The Sociable Weaver, who have been involved in the design and construction of housing and a community which enhances social value called The Cape [18] (see Chapter 9 for more information). These companies will adopt many CSR-related considerations, such as the:

- *innovative design of buildings*, which considers safety during both implementation and use, the social value for end-users, and high energy efficiency and performance. The design should also be conscious of the requirements across the life of the building, with a changing climate and the end-of-life in mind.
- *building materials* should be environmentally friendly and safe, meet legal regulations, and be sourced ethically and locally. There should be close consideration for reducing and eliminating waste during construction and at the end of life.
- *implementation of building structures*, where employees are valued, and nearby residents are considered to avoid disruption through noise and pollution.
- *supporting infrastructure*, with energy sources having reduced and restricted emissions (such as renewable energy), with safe capture and distribution where possible.

- *end-product*, which considers the creation of a lasting quality building structure that is fit for purpose for the design life and then is able to be disassembled and recycled, repurposed, or reused, at the end of life.

The above examples provide some ideas of CSR expectations and goals for organisations constructing within the built environment. It is recommended that to provide a holistic CSR strategy, construction companies should consider both the project life cycle and all of the relevant stakeholders throughout this life cycle. This means the companies need to go beyond picking and choosing CSR initiatives that are perhaps easier to achieve or actions they already do; but instead focus on all potential CSR activities and expectations, with the understanding that the legacy they leave with these projects will impact on communities for decades to come.

CSR actions can be demonstrated even before any construction work has begun by behaving in socially responsible ways during tendering and contracting processes. This includes avoiding unethical practices, including using contract loopholes to make claims or offering bribes to get building permissions or permits quickly. The tendering process presents opportunities for unethical practice that can be difficult to identify, for example, the practice of bid auctioning, where a favoured subcontractor is discreetly returned to (after all bids have been made) and asked to lower their price to match the lowest competitive bids. Other forms of unfair competition include bid rigging from construction cartels. This becomes possible when a group of companies have an oligopoly (limited competition) and where there is an opportunity to collude and artificially raise prices collectively to make excessive profits.

This type of outcome would not be possible in a normal competitive market, as when such agreements are made between organisations that are bidding for work, they eliminate competition. This can be achieved by various unethical practices, such as competitors agreeing not to bid for a specific project, to only bid in a certain geographic region, or to submit a non-competitive cover bid that is too high or contains unacceptable terms. Bid rigging in the construction industry is more common than in other industries since there are often only a few potential bidders, with the tendering process being costly and time-consuming. In some cases, to incentivise bids, losing bidders will be paid a fee for their time, further adding costs to projects. Bid rigging becomes even more possible when there are multiple construction projects that are similar and being constructed around the same time, such as stadiums for major sporting events. This was the case on five stadiums used in the 2014 World Cup in Brazil, where there is evidence emerging that bid rigging occurred, with the second largest construction firm in Brazil admitting to being in the cartel [19].

Following awarding of contracts, construction phase practices should also be undertaken with high levels of consideration for CSR. This is the case for all structures within the built environment that may have slightly different CSR challenges. For example, when constructing new buildings in high-density cities there could be greater consideration for noise and pollution required, or when constructing a bridge over water, there may be greater worker safety considerations needed. It also should be recognised that the built environment goes far beyond the immediate buildings and road connections, but also in the supporting infrastructure with energy sources required to provide essential power with lighting, heating, and other demands, not only at the time of construction but ongoing across the life of the building. This need for energy can result in offshore drilling, nuclear powerplants, gas pipework, and other structures that also can carry further environmental and safety risks.

Sadly, there are many high-profile examples where the extraction and transfer of such energy has been mismanaged, with devasting consequences. The Chernobyl disaster in 1986 is a clear example of catastrophic damage to workers, residents, and the surrounding environment. There were 30 workers that died in three months of the Chernobyl disaster, with radiation exposure causing many more deaths since [20]. The radiation threat meant 350,000 residents were forced to evacuate [20] and the disaster left a devastating environmental impact with land, water, and animals exposed to radiation. More recently, The Deepwater Horizon disaster, which occurred in 2010, is another example of catastrophic safety and environmental consequences for organisations that show a complete lack of CSR. This disaster demonstrates the financial, environmental, and human costs that can occur from a business model that aims to maximise profits without conforming to the basic expectations of society (see Case Study 5.1).

Case Study 5.1   Deepwater Horizon.

In 2009, BP stated in their annual review that: 'Our priorities have remained absolutely consistent – safety, people and performance' [21]. This was despite the criticism and crises they had faced in recent prior years with the Texas City Refinery explosion in 2005 [22] and the Prudhoe Bay spill in Alaska in 2006 [23]. There were 15 workers that lost their lives in Texas and 180 others injured. Houses within a mile of the explosion were damaged and 43,000 residents were ordered to remain indoors. BP had targeted a 25% budget reduction in 2005, which left the refinery's infrastructure and equipment in disrepair while being operated by a downsized and untrained workforce. The US Occupational Safety and Health Administration found 301 egregious violations in their investigation and issued their largest fine in their 35-year history of US$21 billion. A year later an oil spill was discovered in Alaska of up to 267,000 US gallons – which was the largest oil spill ever on Alaska's North slope [23]. The investigation found that BP knowingly neglected corroding pipes and was fined US$20 million [24]. These were two examples of what was to follow in 2010, when the Deepwater Horizon disaster occurred. On the 20th of April 2010, an explosion on the semi-submerged offshore drilling rig named Deepwater Horizon killed 11 workers and ignited a fireball that was reportedly visible from 35 miles away [25]. The inextinguishable fire led to the sinking of the rig and the largest marine oil spill in history [26].

There was a clear paradox between the public image that BP tried to portray and the harsh reality that can be evidenced through their dismal safety and environmental records. In 2001, BP infamously re-branded itself 'beyond petroleum' from 'British petroleum' and pledged to keep emissions constant [27]. BP spent over US$200 million on advertising in an attempt to promote an image as a leader in CSR [28]. They abandoned the rebrand following the Deepwater Horizon disaster. The focus on profits and shareholder dividends meant BP had a complete lack of social responsibility for its employees, the environment, the nearby residents, and other relevant stakeholders. This is a case that shows the consequences of a narrow focus on profits only, with disastrous consequences. If there is to be a positive systemic change from this disaster, it should include lasting amendments to corporate law and practice that encourage companies to engage in genuine, more substantive forms of CSR [28].

The Deepwater Horizon case example demonstrates the devastating impact that a lack of CSR can have. While many companies may have a claim to have good CSR by highlighting one-off initiatives they undertake, such as holding an event with the local community or donating to charity, there are few that have a holistic CSR strategy.

It is proposed that construction companies need to start considering in greater detail how high levels of CSR can be achieved with every stakeholder consistently across the life cycle of the project. Construction companies interact with many stakeholders including partners, suppliers, shareholders, the local community, government, their own employees, and the end-user. For the built environment to become a more socially responsible place, each of these stakeholder interactions requires much further attention. While most companies will demonstrate some form of CSR with each of these stakeholders, there is room for improvement. For instance, when considering the government as a stakeholder, private companies should demonstrate some form of CSR by abiding by regulations, paying taxes, and providing employment opportunities. It is hoped that most companies would show these more fundamental forms of CSR. However, companies could also go further by, for example, supporting public welfare activities the government have initiated and by proactively raising corruption concerns that they become aware of. As previously highlighted, CSR goes beyond confirming to legal government requirements, and construction companies should also consider stakeholders that contribute to project delivery, such as employees, shareholders, partners, and suppliers. A few examples of CSR actions and expectations are suggested below [29]:

- For their own employees, it is expected that they provide a safe working environment, have legal working hours and equal job opportunities, guarantee wages on time, provide opportunities for professional development, and allow workers freedom to join unions or make complaints.

- For shareholders, there should be accurate disclosure of corporate developments, attempts to increase share value, and involvement of shareholders in relevant decision-making processes.
- For partners and suppliers, the contractual obligations should be met in a timely manner, there should be a clear disclosure of policies and expectations, and CSR promotion through recording and assessing CSR commitment.

As well as those involved in project delivery, construction companies should also demonstrate social responsibility for the conservation of energy and resources, protecting the environment, and considering the local community. This could include: maintaining communication with the community on project developments, providing work opportunities for locals or minorities, protecting them from hazards, building community welfare facilities, participating in community activities, and providing some charitable donations where appropriate [29]. Another important stakeholder for construction companies within the built environment to consider is the consumer, which is further discussed in detail below.

5.3 CSR considerations for the consumer

Within the built environment, structures are constructed for both consumers and end-users. These terms can have slightly different meanings. For example, in the case of transportation networks, stadiums, and other structures used by the public, the term 'end-user' is most appropriate. Though when structures are purchased, typically in the form of home ownership, then 'consumer' is most suitable. Consumers of the built environment typically invest large sums of finance, often in combination with significant bank loans, to pay for the cost of their home. With such a large financial investment, consumers of the built environment rightfully expect the product meets all legal requirements, is constructed of durable materials, and provides a safe place to live, that is, free from unacceptable risks to their health. If it is a new-residential build, there is also an expectation that the chosen builder discloses accurate information on their past performance and their credibility: that they complete the project on time, within the budget, and to the needs and expectations of the customer. Where possible, innovative and high-quality materials should be sought, and those materials that are selected for use in the design should not be substituted for cheaper choices later down the project timeline (without approvals and assurance that any substitute still meets minimum building regulation, safety, or performance requirements). The Grenfell Tower fire in the United Kingdom is a recent example where building materials did not reach the basic expectations of consumers. This complete lack of CSR contributed to disaster, as discussed in Case Study 5.2.

Case Study 5.2   Grenfell Tower.

The Grenfell disaster occurred on the 14th of June 2017. A building fire killed 72 people and injured 70 others. In the United Kingdom, this was the deadliest structural fire disaster since the Piper Alpha oil platform in 1988 and the largest residential fatal fire since World War II. The fire spread quickly up the façade of the building in the early hours of the morning, due to the combustible cladding that had been used [30]. The original zinc cladding proposed in design was substituted with the more flammable aluminium composite cladding, saving nearly £300,000 [31].

At the time of writing, the second phase of the inquiry into the disaster had been delayed due to the coronavirus pandemic. Currently there are over 400 hours of evidence from more than 50 witnesses. As the inquiry unfolds, there have been some shocking acts of unethical and dishonest company practices that go against all principles of CSR. The Housing Secretary, Robert Jenrick acknowledged the inquiry has revealed that there has been: 'dishonest practice by some manufacturers ... including deliberate attempts to game the system and rig the results of safety tests'.

Thus far the inquiry has placed intense scrutiny on the companies involved in manufacturing, testing, and selling combustible materials. Arconic, a manufacturer of the cladding used on Grenfell, did not inform various customers, and the United Kingdom body that certifies materials, of their product rating. Their product was subsequently found to fail the safety standards that the manufacturer claimed [32]. Another company that produced most of the insulation on the Grenfell Tower, Celotex, made damming admissions regarding obtaining product approval for its use on high-rise buildings. The company falsified a large-scale fire test (required for use on high-rises) for commercial gain [33]. This was achieved by adding a non-combustible material to the fire test, so the flames would not reach the insulation [33].

Another insulation provider on Grenfell, Kingspan, acknowledged that they sold insulation for high-rise buildings for 14 years without having a credible large-scale fire test. Their insulation had passed a test in 2005, but they subsequently changed the product a year later without retesting [33]. The inquiry revealed that a test in 2007 became a 600°C 'raging inferno', where supervisors had to extinguish the fire as it threatened the laboratory [34]. Kingspan continued to sell the insulation based on the out-of-date 2005 test results until 2020, when the product was withdrawn. A Kingspan executive stated in an email that customers concerned with product safety could 'go fuck themselves' [35]. Directors of Kingspan also sold £6.5m worth of shares before damming evidence was revealed in the inquiry, which has infuriated survivors and the bereaved [36]. A legal representative of a group of victims, Stephanie Barwise, stated that: 'Kingspan's unrepentant arrogance is truly chilling'.

The case continues.

The Grenfell Tower inquiry was formally set up on the 15th of August 2017 and, at the time of writing, is still yet to finish. The 2014 Lacrosse building fire in Melbourne is another example of a complex case which took years to resolve and where there were challenges identifying accountability (see Chapter 2 for more details).

For CSR, consumers should have access to processes and services to manage any customer complaints post-construction [29]. However, these issues with post-construction customer service and consumer protection within the built environment can be highlighted through our combustible cladding research [37–39]. For example, one homeowner who had been involved in trying to chase the builder to fix the combustible cladding on their building stated in a research interview:

> So you think, 'Oh, I don't have to get it all inspected and everything because it's covered by warranty'. And then find out that the warranty is worth very little, and you have to then deal with the builder who's engaged in dishonest practices essentially.

Other homeowners explained that their original builders were not being responsive:

> So the builder has been contacted by the owners corporation manager on several occasions. And we've heard nothing back from them. They were as I understand it, before I was living here, they were a bit more responsive in the beginning, and there have just been regular apartment building post-build issues. But then over time they've became less and less responsive, and based on what I'm being told, it appears that they don't really respond to requests from the owners corporation manager now.

Homeowners felt powerless, if builders decided to not respond, as another homeowner explained:

> They [builders] probably know that if we wanted to get action out of them by the courts, it would cost us more than it would to actually have the issue fixed. So perhaps it's on their part, that they won't respond to issues unless they know that if we actually pursue them through legal action … The other one is that maybe they just can't really be bothered … For the most part there's no real system that can tell them to respond to these sort of cases either. So, the regulator is relatively toothless in this regard unfortunately as well.

The homeowners did not know where to turn. While some had an opinion on who was responsible, many did not know who should be accountable to fix the combustible cladding or how to even get those responsible for taking appropriate actions to address the issue. Another homeowner put it as:

> In this building somebody has somehow approved or arranged for this shonky cladding to go on. Whoever it turns out to be, and I'm sure you have heard this ad nauseam from all these interviews, the one people who have nothing to do with it, are the owners, and they're the ones that are getting pinged for it.

Without a clear direction on who should be fixing the cladding, the owners were often left with the bill. Many felt this was unjust and unfair, especially when their building was often still under warranty. This meant that homeowners and owners corporations had considered whether legal action was appropriate. However, there were financial risks with this approach, as well as the likelihood that rectification work would be delayed while the legal process unfolded. Despite this, some did decide to pursue legal pathways, and even the initiation of this process came at a cost. For instance, one homeowner described their experience as:

> So we engaged him [lawyer] to take the builder on for cladding. Spent just a small amount, two and a half thousand or something. Which was just to send him some letters and all that, which we got a response eventually, that said: 'Hey, we'll look into it.' which means, he's just kind of doing nothing. There was no time frame on what he was going to do, and he didn't look into it. So, then we had to go to the domestic builders' dispute tribunal. And they tried to conciliate but when they came back to us – that there was another wasted three or four months – because they finally wrote to us and said, 'Well, we can't do any conciliation with him.' And I said, 'Oh, why?' Because he [builder] hasn't responded.

There were clear challenges with getting a builder's response, even in cases when legal action was pursued. Homeowners' also expressed other concerns when pursuing legal action, particularly around the issue of builders removing assets or dissolving their company to protect themselves from any potential upcoming legal dispute. One homeowner stated:

> Because all that time you're doing that [legal action], the builders actually knows that you've got to clear the building and he can just start getting rid of all of these assets, if you go to sue him. It's just giving him [the builder] a heads up.

The process of creating a new company to continue the business of an existing company that has been deliberately liquidated to avoid outstanding payments is known as illegal 'phoenixing'. The consumer protection problems associated with phoenixing has led to new legislation in Australia: the Treasury Laws Amendment (Combating Illegal Phoenixing) Act 2020 [40]. This is not the only action the Australian government have been undertaking (in various states) to further protect the consumer and move towards solving the lack of consumer care from the construction industry, as is discussed in more detail below.

5.4 Next steps: Greater consumer protection required as a CSR expectation

Consumer protection is a basic expectation required as a foundation towards establishing CSR within the built environment. The case of combustible cladding has shown that consumers have not been adequately protected, with construction companies refusing to respond and fix defective cladding. The government in the State of Victoria (Australia) was the first in the world to support homeowners financially when they announced a release of AU$600 million to help them rectify buildings with higher risk cladding [41]. The scheme is a step in the right direction and should help address some of the financial impacts for those owners in the highest risk apartment buildings. Other Victorian Government policy actions include the banning of some flammable materials, involvement in the rectification process of higher risk cladding, and a levy introduced to recoup rectification works from industry [42]. At the time of writing, there is also a review of Victorian building policy that has been initiated, in part, to determine how to better protect consumers in the future [43].

The issues with combustible cladding extend beyond immediate rectification of defective buildings but also wider considerations for solving where the problems have manifested from. In New South Wales (Australia), a focus on the future has led to quick and fundamental policy reform aimed at addressing performance on several fronts. The New South Wales reform [44, 45] forms part of the response to 2018 Shergold-Weir Report [46], which found fundamental systemic issues with the building sector. This reform seeks to change the building industry through:

- *improving the regulatory framework* with new powers, processes, and audit practice for the regulator;
- *building risk rating systems* that link past practice, finance, and insurance;
- *improving procurement methods* with major changes to contracting, declared design requirements, and sign off processes/stages;
- *building skills and capabilities* through professional education, development, responsibilities, and certification; and
- *developing a digital future*, where digital systems modernise and harmonise the industry.

A range of supporting legislative reforms were developed through the Department of Customer Service, which have now passed through parliament to drive better building outcomes and in turn better consumer outcome [44]. For example, the Design & Building Practitioners Bill 2020 provides owners corporations with a new retrospective duty of care on defects that extends back ten years. These moves in New South Wales are also promising in this regard for future projects, in terms of reforming policy, practice, and policing with consumer risk mitigation, protection, and confidence in mind.

The effectiveness of such government action is yet to be determined, but it appears to be a step in the right direction for protecting consumers. While it is a basic expectation that consumers are protected through building warranties, these policies have not been adequate in many cases, including that of combustible cladding. More robust consumer protection is required, but this alone will not ensure construction companies will demonstrate high levels of CSR. It is merely one aspect relevant to one stakeholder (the consumer), and there are a range of other CSR expectations that consumers and other stakeholders have through the life cycle of the construction projects undertaken.

There are also other avenues developing which could help reshape the wider construction industry. B Corps, mentioned earlier in this chapter, present one such option [3–5, 16]. There are an increasing number of companies involved in the design, construction, and maintenance of the built environment who have been certified B Corp. They offer a clear difference in how to do business which is good for people, the planet, and financial outcomes and gives the wider construction industry a framework which they could follow and implement to improve CSR outcomes.

5.5 Conclusions

Corporate social responsibility, or CSR, is about organisations being socially responsible for all the stakeholders they interface with. Within the built environment, end-users and consumers are important stakeholders that should be carefully considered during the formulation of a holistic CSR company strategy. Consumers of homes make a significant financial investment and they deserve, at the very least, the basic levels of CSR that society would expect. Examples of these expectations include: that it has been constructed to all relevant codes, standards, and other legal requirements, there have not been shortcuts made through the construction process that reduce the quality of the building, and it is a safe and healthy place to live. The consumer can also expect that the builder discloses accurate information, completes the project on time and within the agreed budget.

Post-construction is where many issues can emerge, and homeowners that are within the building warranty period require clearer paths towards fixing building defects. The case of combustible cladding has highlighted how many builders are unresponsive to high-profile defects once the building is complete, showing a lack of CSR. Governments in Australia are implementing policy to try and address these issues, but the effectiveness of these new approaches is yet to be fully understood. Many construction companies that create the built world are clearly still struggling to fulfil basic CSR expectations – such as fixing defects within warranty periods. In some cases, the lack of clarity on who within the construction supply chain is accountable creates a complex blame game between relevant parties, with no rectification work being undertaken unless ordered by the courts. The legal option is one that some consumers have considered but is not without financial risks and long delays to rectification work. These types of basic CSR expectations need to be met before the wider construction industry can be considered to be socially responsible. Those construction companies that are focused on demonstrating high levels of CSR should attempt a holistic strategy by analysing in depth and detail how to deliver CSR initiatives to all relevant stakeholders across the life cycle of their projects.

References

1. Duman, D., H. Giritli, and P. McDermott, *Corporate social responsibility in the construction industry.* Built Environment Project and Asset Management, 2016. **6**(2): pp. 218–231.
2. Friedman, M., *A Friedman doctrine – The social responsibility of business is to increase its profits.* 1970, The New York Times, 06/07/2021 https://www.nytimes.com/1970/09/13/archives/a-friedman-doctrine-the-social-responsibility-of-business-is-to.html
3. Diez-Busto, E., L. Sanchez-Ruiz, and A. Fernandez-Laviada, *The B Corp movement: A systematic literature review.* Sustainability, 2021. **13**(5): p. 2508.
4. Marquis, C., *Better business: How the B Corp movement is remaking capitalism.* 2020, Yale University Press.
5. Honeyman, R., and T. Jana, *The B Corp handbook: How you can use business as a force for good.* 2019, Berrett-Koehler Publishers.
6. Loosemore, M., and B. Lim, *Linking corporate social responsibility and organizational performance in the construction industry.* Construction Management and Economics, 2017. **35**(3): pp. 90–105.
7. Loosemore, M., and F. Phau, *Responsible corporate strategy in the construction industry: Doing the right thing?* 2011, London: Routledge.
8. Wang, S., *Chinese Strategic decision-making on CSR, CSR, sustainability, ethics & governance.* 2005, Berlin, Heidelberg: Springer-Verlag.
9. Watts, G., A. Dainty, and S. Fernie, *Making sense of CSR in construction: Do contractor and client perceptions align?* In: A. Raiden and E. Aboagye-Nimo, eds. *Proceedings 31st Annual ARCOM Conference, 7–9 September,* 2015. Lincoln: Association of Researchers in Construction Management. pp. 197–206.
10. Trautrims, A., S. Gold, A. Touboulic, C. Emberson, and H. Carter, *The UK construction and facilities management sector's response to the modern slavery act: An intra-industry initiative against modern slavery.* Business Strategy & Development, 2021. **4**(3): pp. 279–293.
11. The Chartered Institute of Building, *Modern slavery: The dark side of construction.* 2015, London: The Chartered Institute of Building.
12. Kärnä, S., *Analysing customer satisfaction and quality in construction – The case of public and private customers.* Nordic Journal of Surveying and Real Estate Research, 2004. **2**: pp. 67–80.
13. Forcada, N., M. Macarulla, and P. Love, *Assessment of residential defects at post-handover.* Journal of Construction Engineering and Management, 2013. **139**(4): pp. 372–378.
14. Ahiaga-Dagbui, D., P. Love, S. Smith, and F. Ackermann, *Toward a systemic view to cost overrun causation in infrastructure projects: A review and implications for research.* Project Management Journal, 2017. **48**(2): pp. 88–98.
15. Xue, X., R. Zhang, R. Yang, and J. Dai, *Innovation in construction: A critical review and future research.* International Journal of Innovation Science, 2014. **6**(2): pp. 111–125.
16. B Corp. *About B Corps.* 2021, 41/11/2021 https://bcorporation.net/about-b-corps
17. B Corp. *B Corp directory.* 2021, 41/11/2021 https://bcorporation.net/directory
18. Moore, T., N. Willand, S. Holdsworth, S. Berry, D. Whaley, G. Sheriff, A. Ambrose, and L. Dixon, *Evaluating the cape: pre and post occupancy evaluation update January 2020.* 2020, Melbourne: RMIT University and Renew. https://renew.org.au/wp-content/uploads/2020/01/Evaluating-The-Cape-research-RMIT_Renew-January-2020.pdf

19. Mano, A., *Brazil builder admits to World Cup stadium cartel in deal with regulator.* 2016, Reuters, 19/01/2021 https://www.reuters.com/article/us-ag-participacoes-corruption/brazil-builder-admits-to-world-cup-stadium-cartel-in-deal-with-regulator-idUSKBN13U2RL?edition-redirect=in

20. World Nuclear Association, *Chernobyl Accident 1986 (Updated April 2020).* 2020, 01/06/2021 https://www.world-nuclear.org/information-library/safety-and-security/safety-of-plants/chernobyl-accident.aspx

21. BP, *BP annual review 2009.* 2009, London: BP.

22. U.S. Chemical Safety and Hazard Investigation Board, *Investigation report – Refinery fire and explosion and fire. Report no. 2005-04-I-TX.* 2005, Texas. file:///C:/Users/e67591/Downloads/CSBFinalReportBP.pdf

23. BBC News, *Alaska hit by 'massive' oil spill.* 2006, BBC, 01/05/2021 http://news.bbc.co.uk/2/hi/americas/4795866.stm

24. Bradner, T., *BP pleads guilty, fined $20M in clean water act violation.* Alaska Journal of Commerce, 2007. **December**(1). https://www.alaskajournal.com/community/2007-12-09/bp-pleads-guilty-fined-20m-clean-water-act-violation

25. Crittenden, G., *Understanding the initial Deepwater Horizon fire.* 2010, Daily News, 05/01/2021 https://web.archive.org/web/20100620121041/http://www.hazmatmag.com/issues/story.aspx?aid=1000370689

26. Republic of the Marshall Islands, *Deepwater horizon marine casualty investigation report.* 2011, Virginia: Office of the Maritime Administrator.

27. Carpenter, S., *After abandoned 'beyond petroleum' re-brand, BP's new renewables push has teeth.* 2020, Forbes, 01/05/2021 https://www.forbes.com/sites/mitsubishiheavyindustries/2020/12/08/the-winds-of-change-are-blowing-further-offshore–with-floating-turbines/?sh=342a0c3cd163

28. Cherry, M., and J. Sneirson, *Beyond profit: Rethinking corporate social responsibility and greenwashing after the BP oil disaster.* Tulane Law Review, 2011. **85**(4): pp. 983–1038.

29. Zhao, Z., X. Zhao, K. Davidson, and J. Zuo, *A corporate social responsibility indicator system for construction enterprises.* Journal of Cleaner Production, 2012. **29–30**: pp. 277–289.

30. Moore-Bick, M., *Grenfell Tower inquiry: Phase 1 report overview, report of the public inquiry into the fire at Grenfell Tower on 14 June 2017.* 2019, London: APS Group on behalf of the Controller of Her Majesty's Stationery Office.

31. Symonds, T., and D. De Simone, *Grenfell Tower: Cladding 'changed to cheaper version'.* 2017, BBC News, 01/20/2021 https://www.bbc.com/news/uk-40453054

32. Symonds, T., and C. Ellison, *Grenfell Tower cladding failed to meet standard.* 2018, BBC News, 01/20/2021 https://www.bbc.com/news/uk-43558186

33. Lamble, K., *Grenfell Tower inquiry: 11 Key things we have learned this year.* 2020, BBC News, 01/20/2021 https://www.bbc.com/news/uk-55349395

34. Booth, R., *Grenfell firm stretched truth with fire safety claims, ex worker says.* 2020, The Guardian, 01/20/2021 https://www.theguardian.com/uk-news/2020/nov/23/grenfell-firm-stretched-truth-with-fire-safety-claims-ex-worker-says

35. Booth, R., *New watchdog will be able to ban dangerous materials used at Grenfell Tower.* 2021, The Guardian, 01/20/2021 https://www.theguardian.com/uk-news/2021/jan/19/new-watchdog-will-be-able-to-ban-dangerous-materials-used-at-grenfell-tower

36. Booth, R., *Director of Grenfell firm quits after evidence it used outdated fire tests.* 2020, The Guardian, 1/20/2021 https://www.theguardian.com/uk-news/2020/dec/17/director-grenfell-firm-quits-evidence-outdated-fire-tests-kingspan

37. Oswald, D., T. Moore, and S. Lockrey, *Combustible costs! Financial implications of flammable cladding for homeowners.* International Journal of Housing Policy, 2021. pp. 1–21. DOI: 10.1080/19491247.2021.1893119.
38. Oswald, D., T. Moore, and S. Lockrey, *Flammable cladding and the effects on homeowner well-being.* Housing Studies, 2021. pp. 1–20. DOI: 10.1080/02673037.2021.1887458.
39. Oswald, D., *Homeowner vulnerability in residential buildings with flammable cladding.* Safety Science, 2021. **136**. DOI: 10.1016/j.ssci.2021.105185.
40. Parliament of Australia, *Treasury Laws Amendment (Combating Illegal Phoenixing) Act 2020.* 2020, Canberra: Federal Register of Legislation. https://www.aph.gov.au/Parliamentary_Business/Bills_Legislation/Bills_Search_Results/Result?bId=r6325
41. Andrew, D., *Tackling high-risk cladding to keep Victorians safe.* 2019, Media Release, Victorian Government, 11/20/2019 https://www.premier.vic.gov.au/wp-content/uploads/2019/07/190716-Tackling-High-Risk-Cladding-To-Keep-Victorians-Safe.pdf
42. Thorburn, M., *Cladding's burning issue.* Law Institute Journal, 2020. **June**: p. 60.
43. State of Victoria Government, *Expert panel on building reform: Terms of reference.* 2020, State of Victoria Government, 01/25/2021 https://www.vic.gov.au/building-system-review
44. NSW Building Commissioner, *Building reforms boosted with new transformation team.* 2020, Department of Customer Service, 08/10/2020 https://www.nsw.gov.au/customer-service/news-and-events/news/building-reforms-boosted-new-transformation-team
45. NSW Government, *Rebuilding confidence.* 2020, Department of Customer Service, 08/10/2020 https://www.nsw.gov.au/building-commissioner/rebuilding-confidence
46. Shergold, P., and B. Weir, *Building confidence – Improving the effectiveness of compliance and enforcement systems for the building and construction industry across Australia.* 2018, Canberra: The Building Ministers Forum.

6   Dealing with dangerous defects in multiple occupancy developments

Key takeaways

- There has been an increase in multiple occupancy buildings around the world. This type of housing involves individual dwelling ownership alongside collective ownership of common property, resulting in the need for group decision-making and consensus when responding to building issues such as defects.
- Professional strata managers have a critical role to play in order to create social value in multiple occupancy buildings. This includes proposing sustainable and social initiatives that improve both the quality of buildings and engagement within owners corporations.
- Professional strata managers and owners corporations deal with common issues that emerge in multiple occupancy developments. However, research shows they require further support for dangerous defects, such as flammable cladding, as these significant and complex defects go beyond the skillsets of those involved.
- Further government support structures are required to help individual consumers collectively navigate dangerous defects within multiple occupancy buildings.

Chapter summary

This chapter focuses on residential properties where there are multiple units on a single piece of land. These units are owned individually, but the common area of the building is owned collectively by all the owners. Multi-unit developments often refer to apartments (or similar such as condominiums). However, multi-unit developments can also refer to lower density housing estates where houses are owned individually but there are areas of common property across the estate, such as is seen throughout areas of North America. In this chapter, we discuss multi-unit dwellings in terms of apartment developments; however, the points raised are relevant across a range of multi-unit types. The chapter provides an overview of the management of multiple occupancy buildings through groups of owners (e.g. owners corporation) and supporting strata management professionals. The chapter also explores the challenges that multiple occupancy buildings have when

DOI: 10.1201/9781003176336-6

responding to dangerous defects, such as flammable cladding. There are suggestions to improve the social value placed on the building that attempts to increase owners engagement in building issues, improve building quality and performance, and improve the residential experiences overall. While owners corporations and professional strata managers provide a good structure for addressing many issues which may arise with buildings, for dangerous defects, such as flammable cladding or asbestos, support from the government will likely be required, due to the significance, complexity, and cost of rectifying severe defects.

6.1 An introduction to multiple occupancy developments

Around the world, there has been a shift towards higher density living in multiple occupancy developments [1–4]. The broader shift towards densification globally is being driven by factors including affordability, location, and sustainability [5–8]. This has led to more groups of owners within a building (or wider development) that own their dwelling but share the common property with other owners. Some jurisdictions have a long history with higher density living, particularly throughout regions of Europe. Others, like the United States of America and Australia, are comparatively new to this type of housing and living. For example, in the United States of America there has been an increase of individual dwelling ownership with the shared common property as a percentage of the total population from 1% in 1970 to more than 23% by 2019 [9]. In Australia, the percentage of apartments of the total housing stock increased from ~12% in 1971 [10] to more than 27% in 2016 [11]. The global increase of higher density living has implications for the way cities and housing is designed, constructed, and experienced, including the different tensions and challenges that living in common housing can bring [1, 12–23]. This, in turn, can have a significant impact on creating and establishing greater social value within multiple occupancy buildings, particularly when issues of defects and quality arise.

The consumers within multiple occupancy buildings typically have the collective responsibility to maintain the shared common property [12–16, 24–28]. Their effectiveness in maintaining the common spaces, such as entrance ways, stairways, gardens, gyms, and pools, is an example of where social value can be positively or negatively created within these buildings. The multi-unit consumer group has different opportunities and challenges for creating social value than the individual consumer living in detached on semi-detached housing [29–34]. It is a form of ownership that can be more complex due to the need for group decision-making and consensus across a range of diverse household needs and requirements. As Easthope [13] states:

> [a]partment living requires that we try to get along with the people who live close to us in our building, but we also need to make decisions collectively about how the building we live in will be managed. What kind of a place do we want it to be? What behaviours are acceptable amongst our neighbours? How much do we want to spend on maintaining the building, and do we want to make any improvements? Occasionally everyone will agree, but most of the time they won't.

These questions can be particularly challenging for consumers to successfully answer since the social relationships and connections between building residents may not be established. While multiple occupancy buildings do provide an opportunity for high social engagement, establishing social relationships and connections can be difficult because:

- It is likely that new purchasers (or tenants) will know few, or even any, residents within the building.
- It may not be possible to predetermine the types of people that occupants may be sharing the building with.
- It can be difficult to create social relationships once residing in the building due to building design.
- Residential living in multiple occupancy buildings can be transient, as people often move in and out of buildings within shorter time horizons than detached housing.
- There can be a high percentage of investor-owners who own apartments resulting in owners who are detached from the social fabric of the building, since they never plan to live in the building and may never even have visited the building. They are also likely more focused on financial, than social, outcomes.
- Purchasers may be drawn to live in higher-density housing for security in numbers, without needing to know their neighbours.

The various social dynamics within multiple occupancy buildings can make it challenging to manage such buildings and occupants, as collective ownership decisions are often required for any changes to the building or governance of the building. Often, the lack of active contribution to the management of the building from occupants has wider-reaching impacts for social value. Part of the challenge to improving active engagement with the owners corporation (see next section) and decision-making is that many people in apartments have been found not to know their neighbours very well at all [15, 24]. In one study, two participants who had lived on the same floor for 30 years said they barely knew each other [35]! The way many multi-unit developments have been designed means there are often not those natural opportunities for casual interactions and the frequent turnover of residents in some buildings contributes to weak social ties between occupants [36]. This makes it increasingly difficult for many owners corporations to have strong engagement from all residents. This, in turn, means managing multiple occupancy buildings through collective decision-making is very challenging.

6.2 The management of multiple occupancy developments

In order to manage multiple owners, units, and owner/occupant interests across a building, different governance measures have been developed around the world. One particular mechanism has been the development of owners corporations, and by extension strata managers, to manage individual ownership rights within the collective responsibility of the building or development [13, 24, 27]. Owners corporations have different names around the world, such as commonhold, homeowner/resident associations, housing and condominium cooperatives, and owners governance, to name a few. Despite the various names, they tend to have a similar intent to how they operate and what they do. In this chapter, we will refer to this type of building governance as owners corporations.

An owners corporation is essentially a building management structure and is often a legal requirement to have in place when there are multiple dwellings on a single parcel of land or where there is significant shared common property. When you become an owner of a unit in a multi-unit building or development, this automatically makes the owner a member of the owners corporation. They are often managed by a smaller committee of owners who look after the day-to-day management of the building, with larger decisions put to all owners for consensus decision-making (such as decisions with certain financial parameters, making changes to the way the common property is used, and major repairs). The owners corporation committee is often, but not always, managed by a professional strata manager,[1] who manages the day-to-day affairs for all owners and tenants. They coordinate with the committee, arrange repairs, collect levies, manage finances and insurance, and handle complaints or issues. They have the opportunity to support owners in a variety of ways, such as:

- providing an explanation of legal rights and responsibilities of being an owner;
- gathering evidence and supporting owners in building insurance claims (e.g. for defects);
- taking steps towards sustainable initiatives, such as LED lights, electric vehicle charging, solar panels, and rainwater tanks;
- proposing ways to improve food production and biodiversity through the use of vegetable patches, plants, and small gardens;
- proposing ways to reduce risks of building graffiti, with use of trees, plants, or street art; and
- considering purchasing goods and services from the local community.

Some of these initiatives, such as proposing the investment for solar panels, go above and beyond normal day-to-day affairs and expectations. However, for these initiatives to manifest, they will often require a strata manager to start the process and have answers to concerns that homeowners may have. These could be on the implications solar panels would have for roof warranty or potential water leaks, as well as how it would be funded and how benefits would be spread across occupants. A strata manager can demonstrate going above and beyond the usual day-to-day building operations, by considering initiatives for the benefit of homeowners. There are many benefits to improving the quality of the building (that the strata manager can instigate in the post-handover stage). These benefits include property value, sustainability, and social value since homeowners will typically value their homes more upon improvements. Further, social value can also be created through the process of proposing, considering, and agreeing to make building improvements including proactive maintenance and upgrades.

This is often because new social networks and relationships can be established within the building during this process. Planning and proposals for building improvement are one way to also improve the engagement of building occupants. For example, owners corporation meetings that were perhaps poorly attended can often become well-attended and meaningful when there are considerations for building improvements, such as solar panels, electric vehicle charging, and community gardens.

While these changes can be proposed by the strata manager to begin the process, they ultimately need to be agreed upon by the owners corporation. In some jurisdictions, owners corporations are mandatory for certain types of developments; in others they are voluntary. They also range from being solely unit-owner led to those who engage (either as a legal requirement or voluntarily) a professional building or owners corporation manager to undertake all, or some, of the requirements of the owners corporation.

The purchase of a unit gives the individual owner certain rights about what they can do within their own unit, with the owners corporation responsible for the management and use of the common property (e.g. corridors, lifts, entrance/exit points, the wider grounds, internal roads, parking, and any leisure/fitness facilities). However, in many jurisdictions the owners corporation can also establish and enforce rules or requirements for how individuals use their units. This is if/where activities impact neighbours, such as constraints around noise and having pets, within the building. The specific function of an owners corporation varies around the world but typically has been introduced as a mandatory requirement by governments to undertake [24, 25, 28], including:

- management of any common property in the building,
- conducting building maintenance on common property and services,
- planning for and coordinating major capital works,
- facilitating emergency repairs and addressing building defects,
- taking out appropriate building and common property liability insurance,
- collecting and managing fees from unit owners,
- creating any by-laws governing what occupants can or cannot do within the building and their unit,
- addressing any breaches of owners corporation rules,
- managing communication and organising regular committee meetings,
- undertaking dispute resolution, and
- adhering to any legal requirements of the jurisdiction.

There are requirements, or at least recommendations, in many jurisdictions around the world that owners corporations engage a professional strata manager to oversee many of these above activities. For example, in Australia a professional strata manager is typically appointed by the owners corporation and provided a clear remit for the work they will undertake. This approach helps ease the burdens of property management for owners in the building

who often do not have the experience or knowledge of how to manage a property or the wider socio-legal requirements the building occupants might need
to adhere to. As such, a professional strata manager could be a useful resource
when building defects emerge, and owners lack understanding and engagement to organise rectification. Though it is important to note that while a
strata manager may be engaged, this does not diminish the requirement that
individual owners contribute via the owners corporation and, in particular,
key decision-making.

The advantages of having an owners corporation (beyond any legal
requirement to have one) and strata manager includes that the costs for the
management, security, maintenance, and upkeep of the building are shared
amongst all owners. It also provides a structure whereby any individual or
collective issues in the building can be addressed and allows for more systematic and coordinated plans for how to maintain the building quality or
undertake improvements. Thus, while the rationale for using a strata manager
is clear, an ongoing challenge around the world is that strata, building and/or
property management is not a universally regulated or governed industry, and
in some locations there are no minimum education or experience requirements to become a strata, building or strata manager. In the United States of
America, for example, property management has been described as 'an incho-
ate profession where there is enormous variation in the education, training,
experience, wisdom and honesty of practitioners… [leading to] a number of
scandals involving embezzlement of massive amounts of association funds by
management firms' [26]. This is despite the fact that property managers can
be potentially managing multi-million dollar assets and dealing with a range
of complex governance and social structures, elements which are exacerbated
when dealing with defects or issues of quality.

Further, strata managers are increasingly needing to do more than manage the
physical building and finances of the owners cooperation, with people management and conflict resolution skills increasingly part of the role [38–40]. In the
case of building defects, they also may need to help deal with multiple different
stakeholders with constrained time frames, such as the building insurers, original
builders and the owners corporations. While strata managers can provide support in such situations, the day-to-day decision-making and management of the
building and common areas tends to fall to a smaller number of residents who
make up the owners cooperation committee [26].

The ongoing challenge of engaging owners, in both the wider decision-
making and contributing to any formal owners corporation committees,
means that the burden of management is not spread evenly across all owners. Instead, they are left to a few owners, which creates issues when larger
issues like defects emerge. This·is compounded by the fact that owners are
often frustrated or disillusioned with the owners corporation process and so
remove themselves from contributing, further reducing those who actively
contribute [13, 40]. While a lack of all owners contributing to the management of the building may not be a critical issue when everything is going

well with the building and between residents, it becomes problematic when larger issues such as defects need to be addressed [13, 39]. For example, the case of a Downtown Sarasota apartment in the United States of America highlights this issue [41], where following near building collapse, the process of navigating major repair work took several years. When major defects emerged and residents were evacuated, there was still a lack of engagement from residents to the building management association, which was left to a handful of owners to navigate the process for almost five years on behalf of all owners. Another example of delayed rectification work in the United States of America ended in tragedy, with the collapse of a 12-storey building in Miami in 2021 (see Case Study 6.1).

Case Study 6.1   The cost of indecision: The Champlain Tower collapse (United States of America).

Just before 2 am on June 24 2021, the Champlain Towers South Tower in Surfside, South Florida, partially collapsed [42]. The 12-storey building with 136 apartments was built in 1981 on reclaimed wetlands. More than 55 apartments were destroyed in the initial collapse and 98 people lost their lives.

While the specific cause of the collapse was not confirmed at the time of writing this chapter, what was known was that a building inspection audit from 2018 and again in 2020 had identified a number of defects and maintenance that needed to be undertaken in the building. In total, it was estimated that the costs for addressing these issues would be around US$9.1 million [8, 43]. While the engineer did not indicate that the building was in immediate danger, they did warn that the work would need to be undertaken in the near term.

It has been suggested that there was a lack of action towards undertaking these repairs by the condo association. However, the condo association has defended their progress with reports that they had taken out a loan to cover the work and had been moving through the process to start the work. It was suggested that there were delays due to constant pushback by members of the association and other residents about what to focus on and how to apportion costs for the work [44]. It was reported that in the months leading up to the collapse that the price for the repair work had been revised to more than US$15 million, with residents to pay at least US$100,000 each, which was causing infighting about what work was needed [22, 45].

Gregg Schlesinger, an attorney specialising in construction defects and a former construction project engineer, said these were all problems that should have been dealt with quickly [46]:

> The building speaks to us. It is telling us we have a serious problem ... They (building managers) kicked the can down the road. The maintenance was improper. These were all red flags that needed to be addressed. They weren't ... They failed to do what a reasonably prudent condo association board would have done under similar circumstances.

Given many of the recent and emerging building quality defect and crisis issues are being seen more in higher density or multi-unit developments, this has placed owners corporations right in the middle of responding to this issue [12–15]. We explore this for the remainder of the chapter primarily through the lens of cladding crisis, as a way to demonstrate the challenges owners corporations have when trying to navigate an emerging crisis.

6.3 Experiences of owners corporations during a defect crisis

Our own research exploring the flammable cladding crisis in Australia, along with wider research on how owners corporations address defects, maintenance, and other building quality issues, has identified a number of challenges which are heightened during crisis or dealing with major defects [13, 15, 26, 40, 47–49]. These include:

- owners not contributing (either formally or informally) to the management of the building and any decision-making processes that are undertaken;
- owners corporation committee members and/or the strata manager having a lack of skills, time commitment, and experience to deal with building defects or the crisis;
- navigating motivations of each unit owner, making it challenging to achieve consensus on major decisions such as how to address defects;
- conflicting on which owners should pay for defects if they do not affect everyone (e.g. cladding coverage on some apartments but not others);
- the challenges of trying to undertake rectification work while the wider process (e.g. government) and support (e.g. financial) is continuing to change;
- obtaining clear and consistent information from experts and stakeholders; and
- the high financial costs of solutions.

The remainder of this section touches on a number of these points which emerged from our research with owners, owners corporation committee members, and strata managers in relation to the flammable cladding crisis.

We found that a key barrier to progress flammable cladding rectification works was partially due to the lack of engagement by all impacted owners within a development. These owners who did engage were often time poor, as they had work and family commitments. They reflected the desire to have other owners help share the workload of tasks for defect rectification. It was also raised as an issue that it was also hard to engage owners of the wider building when it came to formal decision-making, such as needing to vote on what pathway to take (particularly where it related to cost implications). This challenge to reach clear and timely consensus was impacting the safety of those in the building, as building work would be delayed and there became a reliance on short-term interim safety measures (e.g. no smoking on balconies) for the long-term. Several participants in our research reflected that they were concerned that any further delays to addressing safety measures could result in the building receiving an eviction notice for failing to comply with safety requirements. As one respondent states:

> [T]here's only three of us on the committee. It was hard for us to sort of work out, 'Is this sufficient for what we need? Is this going to take into account the needs of the various owners?' … we were trying to get assistance from other owners, but there's very few that helped, or become involved.

Another explained that even as they started going through various legal processes to try and get the developer to pay for rectification costs, there was still little engagement from occupants:

> Since then [legal processes starting], we have had monthly forums [for all owners], which are attended in the building foyer on a Sunday afternoon. It's a set time and they get the same kind of turn out, which is probably one eighth of the residents and the owners.

It was also challenging to communicate requirements to the wider owners in a way they could understand or would prompt them to engage:

> One of our major difficulties is the decision to replace the combustible cladding with non-combustible cladding and to justify that to the other owners, as well as the expense to a naïve [owners] corporate committee.

However, we also found in our research that there were some examples of where there was good engagement from many of the owners who actively contributed to the wider decision-making processes. In these cases, it was clear that wide and timely engagement was important for how the owners corporations were functioning, but also meant they had been able to more easily navigate the flammable cladding rectification process to that point in time:

> I would only say that there is an enormous amount of pressure from other owners to make it happen and get it done … this current committee has been the most, well hardworking and responsible working group that I've been part of, and I know everyone is in favour of getting it, getting it attended to as quickly as possible.

The challenge with engaging owners is more problematic with landlords who do not live in the building (see Chapter 7 for more discussion on landlords). While they are owners, landlords were noted as having competing priorities compared to owner-occupiers resulting in additional tensions between the two types of owners. Specifically, participants in our research highlighted that landlords typically had different views on potential solutions, often looking for the most cost-efficient solution that would not require them to overcapitalise on the outcome; whereas those who were owner-occupiers were often more concerned with ensuring that the safety requirements, were not just met, but exceeded.

It is not only the challenge of having people actively contribute to the management of the building but also that owners corporations have often not had to deal with these types of defects previously and have limited capacity to do so. This meant that owners corporations needed to fill these skills and knowledge gaps by hiring experts such as lawyers. This increases the costs for owners corporations which were then passed onto all owners through owners corporation fees or special levies. This was not helped by the fact that the wider government process in Australia for undertaking rectification work has been developed on the go and has often shifted with little warning to those involved. The following is just some of the respondent reflections on this issue from our research:

> Our owners corporation committee, they are spending an awful lot of their time [addressing the flammable cladding rectification]…from meeting once a quarter, they're meeting every fortnight or even every week, just having to deal with this.
>
> I became the unpaid owners corporation Chair and have spent over 200 hours teaching myself the codes and solutions and reading cases and appeals and finding all our original documents … I have been online many evenings teaching myself how to read building codes and fire engineer reports processes and options. The paid "experts" and "managers" seem to have a conflict of interest.

Our research also highlighted social tensions around who pays and why, the scope of rectification, politics on the committee, and navigating a process which is changed on the run. The issue around who pays and why was raised by participants who were in buildings with uneven flammable cladding coverage as creating significant social tension within buildings. Some residents who did not have flammable cladding directly on their apartment were refusing to contribute to the overall costs of rectification, arguing that their unit was safe. However, others argued that the costs should be shared across all owners regardless of the level of cladding on individual units. This opposition to paying an equal amount was creating significant challenges for owners corporations:

> [T]here is one owner who did object to it actually … She is what you'd call a litigious personality, and she is actually a lawyer by background. She's refused to pay for the costs, her portion of the cost we engaged in for our engineer on the grounds that her unit, is like my unit, out of scope for having cladding replaced.

In our research, participants explained that they were unable to progress with rectification works in their building as some owners had refused to pay their special levies:

> We [the owners corporation] introduced a cladding levy. Some owners refuse or are delaying paying their levy. A project manager company and

a sub-contractor have already been appointed and are waiting for sufficient funds to be on hand before commencing.

Linked to this challenge around who pays for the rectification work, there were also challenges with what path to take. In some buildings, a lower-cost option, such as increasing the number of sprinklers, was thought to be acceptable to deem a building safe which meant the flammable cladding could potentially be left on the building. However, in some of these buildings, some owners were against the 'light touch' approach and were advocating for a full rectification.

Such opposing views created significant tensions for many within buildings. This was challenging for those involved in the owners corporation who were trying not only to represent their own needs as an owner but also to try and work through what is best for the building as a whole. For example, one participant, who was on the owners corporation for their building, reflected:

> [T]hat's also a conflict of interests. If I make a decision which minimizes the cost I have to pay, then it could be argued that I've made that decision, that I've argued for that because it advantages me. So, it actually puts me in a difficult position in a way. It's the same with everyone on the owners corporation committee. So, while there's a lot of people [owners] who are not contributing, it wouldn't surprise me if they suddenly developed strong views about the apportionment [of funds], once they actually find out how much money is involved.

The question of costs was an ongoing point of contention for many owners corporations attempting to navigate their way through the flammable cladding crisis. While some owners are more able to afford unexpected costs for rectification, there were others who reported the additional costs through special levies would push them into financial debt (see Chapter 3 for further discussion). There were questions around cost for owners corporations that were difficult to answer, as there was a lack of certainty about what the costs would be, and any notion of if the costs are reasonable. There was a feeling that the industry was taking advantage of the situation, given the limited number of companies who could do the required rectification work and the necessity of it. As one owners corporation member stated to us:

> We obtained one quote [for cladding rectification] which was so outrageously expensive that we wondered if the contractor had confused this with another building.

The uncertainty of how much the rectification work would cost was prompting some owners corporations to start collecting additional funds in order to spread the impact out for owners:

> The treasurer has put together an amount that he thinks was reasonable … because we're having to make provision for this expense [the flammable cladding] … We don't know when it's going to be. And so just trying to get a

bit in the kitty to, to cover it up our expense, rather than have to let [it be collected in one larger amount].

It was not just the issue of the flammable cladding costs, with uncertainty about the other defects that had been identified once cladding came off, which was an issue.

Our owners corporate were advised that we have to pay to remove this cladding. We have been offered up to AU$1.2 million in assistance [from the government] but this will only be reimbursed [after the work is completed]. We also have a number of other building orders [defects] from [the government] that need to be fulfilled before the cladding can be addressed. All are due to the negligence of the builder.

Our research also revealed that while the owners corporations needed external help from building professionals, they did not know who was trustworthy and competent to help them address the cladding defects. For example, one home-owner stated:

In the fire engineer interim report I found 52 inconsistencies or incorrect assumptions, none of which were addressed or ever answered by the fire engineer. In the final report I found another 30 inconsistencies.

Further, when owners corporations did engage building professionals, they sometimes did not know what to ask for to help resolve the cladding issues. As one research participant reflected:

The owners corporation paid an engineer to test the cladding and did not give the engineer a clear scope of work of what was needed. We paid over AU$6,000 and the information was totally useless. The owners corporation committee have no idea about what to ask for when they engaged the engineer. What a waste of money.

As well as challenges with engaging external advice from building professionals and the government, there were also social dynamics within the owners corporations that were revealed to be problematic. Our research reported that despite multiple requests, there were owners corporation committees that were refusing to pass on relevant information to the wider unit owners. It was often framed by the owners corporation under the issue of confidentiality around the process or that providing all the information would overwhelm owners and create issues with working towards a clear resolution. One participant reflected:

I think that all the committee members should get copies of any reports that are provided as soon as they arrive. But also I think if we've had a report from a specialist, then there needs to be a little newsletter to the other owners, to say these are some of the issues that the committee's working on ... They shouldn't hold that information close to their chest … So when you think that they've had these reports, and it's not being shared, you do think I wonder what else they know about it they're not telling us?.

In one case an owner in the building who was not part of the owners corporation was forced to take legal action against their own owners corporation committee to get the information they had about what was happening to their unit and development.

Our research found that a key issue for owners corporations was often the politics and infighting that occurred between members on the committee. This related not just to differences in circumstances or views on what approach to take for rectification but was often linked to the builder or developer still being on the owners corporation committee where they still owned units in the development. For example, one participant in our research spoke about how they were unable to get their owners corporation to begin the process of cladding rectification because the developer (and their friends) owned several units. This meant the owners corporation committee could not achieve consensus with decision-making, since the developer was (not surprisingly) unwilling to take action that would result in them facing a financial loss. In the end, this participant sold their property for a financial loss because of the negative impacts this was having on their well-being. Other participants faced similar frustrations, with one telling us:

> Our building is governed by a dysfunctional committee, apathetic body corporate and building management scheme that is controlled by the developer who is more interested in protecting their interests, not the safety and best interests of the building and owners and residents.

Our research adds to the mounting evidence which finds that even with clear guidance or rules on how owners corporations are meant to work, there are increasing tensions and disputes created with higher density living. The issues emerging within multiple occupancy buildings have an impact on the social value attached to both individual households and the wider building [12–15, 24, 25].

6.4 Next steps: Considerations for owners corporations and strata professionals

Owners corporations are a group of individual owners that have to navigate complex issues, when issues such as building defects emerge. Our research highlighted a variety of challenges owners corporations experience with combustible cladding, including:

- a lack of engagement from owners to contribute to the owners corporations making it difficult to progress and find consensus;
- difficulty knowing which building consultant or professional to ask for help and what to ask for;

- owners corporations not being forthcoming with information to other owners creating the impression there was something to hide;
- social tensions created from fire safety rule breaking and from some owners refusing to pay for cladding rectification;
- representatives from the developer and/or builder on the owners corporation (if they also owned property in the building), creating a potential conflict of interest;
- the high financial costs of potential solutions and the concern the building industry was taking advantage of the situation to drive up costs;
- that the slow and continuous changes to government processes and support exacerbate issues.

Even when there is support from a professional strata manager, much of the contribution of the owners corporation still relies on volunteers who are '… unpaid, untrained, often unqualified, and almost entirely unsupported …' [26]. This lack of skills, knowledge, and engagement plays out in different ways when dealing with minor and major building defects. For example, Robertson [27] sums up the cascading issues by stating:

> Collectively these problems have a direct bearing on the condition of multio-wned property, especially where there is no active management arrangement in place. In such circumstances crisis management becomes the order of the day, activated by a leaking roof, or perhaps a structural collapse caused by a storm. When contractors are needed in a hurry they are randomly selected and are almost invariably of indeterminate quality. Not surprisingly, such contractors have no knowledge of the block's previous repairs history. Poor workmanship and bad working practices tend to ensure poor value for money. Again, due to a combination of speed and an overriding desire to save money, owners seldom make use of either an architect or surveyor to specify works. As a consequence, proper written tendering procedures are rarely initiated. Work procured in this way tends to alter once on site, with limited options to solve unforeseen problems being presented to owners as and when they arise. Inevitably costs rise, which further generates dissatisfaction and disagreement amongst fellow owners who realise they have no real control over the process.

The discussion above, both from the wider literature as well as our own research, highlights that owners corporations have an important role to play in managing both owners and the outcomes in multi-unit developments when addressing larger scale defect or crisis events. While the wider literature notes challenges around occupant engagement with owners corporations and the tensions that can be created by multi-unit living and governance, our exploration of owners corporations during the flammable cladding crisis finds that these are exacerbated. The reality is that owners corporations are not set up or capable of dealing with a large-scale defect issue or crisis that often goes beyond the skillsets of those involved.

In previous research, Easthope, Randolph, and Judd [25] outlined the following reflections to try to improve occupant satisfaction during major renovation and maintenance work in multi-unit developments, including:

- that there needed to be improved acknowledgement by all owners about their responsibilities as members of an owners corporation, and that these responsibilities relate not just to their individual unit by the common property;
- that outcomes are improved with an active and responsive owners corporation committee and owners corporation manager (where used);
- that improving relationships and flows of information about key works such as major repairs and maintenance between the owners corporation committee, owners corporation manager (where used), owners, and tenants are critical;
- that there must be a regular maintenance schedule and a plan for major capital works developed and updated. Such a plan must draw upon expert independent advice;
- that owners corporations must ensure effective financial planning to fund planned and future required works; and
- that any repairs and maintenance works that are undertaken are properly funded and multiple quotes are received for works before a contractor is chosen and work is undertaken.

Many of these points are relevant for preparing owners and owners corporations for also dealing with more significant defect or crisis events.

We would emphasise that many owners, and those who contribute to the owners corporation committee, often have a limited understanding of what is required to manage a building. Owners should be made aware at the time of purchase of what their rights and responsibilities are and that there is a need for active engagement within the building in relation to managing outcomes. While governments do provide some information as to what people's responsibilities are as an owner in a multi-unit development, it is clear from our research that people either do not understand this or ignore this. Strategies to increase engagement should be a priority and the first step should be improving education around owners rights and responsibilities, but also expectations. Strata managers should be key stakeholders for this education process moving forward.

Strata managers could also help with improving formal and informal engagement within a multi-unit dwelling. This would not only improve responses to issues like maintenance and defects but also improve a range of other social value outcomes. There should be a focus on improving the quality of buildings where possible, such as with sustainable and social initiatives. Strata managers that take the time to propose ways to improve buildings will likely create social value through meaningful discussion at more highly engaged owners corporation meetings, create social value through building social relationships and connections within the building and create social value by improving the building quality for consumers. The more owners are engaged with issues around building quality and maintenance, the more prepared they will be when unexpecting building defects emerge, such as flammable cladding.

Along with the need for improved owners corporation engagement, there needs to be an acknowledgement that clearer structures are needed for helping owners corporations during a crisis-type event. While there are avenues for dispute resolution which range from internal to external processes (in Victoria, Australia, for example, an owner could go to the Victorian Civil and Administrative Tribunal) there may need to be a revision to these processes during crisis, to more quickly respond to and address occupants needs. This would require a fundamental rethink of government support structures for consumer protection, which is further discussed in Chapter 8.

6.5 Conclusions

The shift to higher density living has resulted in a rise of individuals who own or live in a unit which is part of a larger development. In order to navigate the issue of individual ownership as part of a collective with common property mechanisms like owners corporations have been developed. While owners corporations offer a way to manage a building and address elements such as regular maintenance, our research looking into the flammable cladding crisis reveals they are not as well suited to dealing with larger defect or crisis issues. A number of issues were revealed around dynamics (e.g. lack of owner contribution to the owners corporation committee), dilemma (e.g. self-interest compared to best outcomes for the building), dignity (e.g. trying to get and share information), and democracy (e.g. conflicting interest of owners corporation committee members).

For owners corporations to be more prepared for unexpected defects, greater engagement is required within owners corporations. It is suggested that the professional strata managers can play an important role in creating engagement by proposing ways the quality of the building could be improved. This could be, for example, proposing the installation of solar panels for sustainable and financial reasons, improving community gardens with wider biodiversity (e.g. vegetable patches), or creating social sub-committees. These types of initiatives create social value themselves by providing greater opportunities for social engagement while improving the building.

While providing greater education and engagement within owners corporations can lead to better outcomes with responding to defects, improving building quality and the residential living experience, there is a need for considering further support when dangerous defects, such as cladding or asbestos, emerge. These types of defects are largely beyond the capabilities of a group of individual owners, who require support through the process of rectification. More robust consumer protection structures are required for such dangerous defects and are discussed in more detail in Chapter 8.

Note

1. The terms strata, building, and property manager are often used interchangeably but have differences [37]. The building manager is typically stationed on-site, is the point of contact for contractors, and is responsible for building cleaning and maintenance services. They report issues to the strata manager. A property manager usually works for a real estate company and represents the owner and tenants of a property in the building. They collect rent, pay bills, and organise repairs for the individual unit.

References

1. Easthope, H., R. van den Nouwelant, and S. Thompson, *Apartment ownership around the world: Focusing on credible outcomes rather than ideal systems.* Cities, 2020. **97**: pp. 102463. DOI: 10.1016/j.cities.2019.102463.

2. Nethercote, M., *Melbourne's vertical expansion and the political economies of high-rise residential development.* Urban Studies, 2019. **56**(16): pp. 3394–3414. DOI: 10.1177/0042098018817225.

3. Moore, T., R. Horne, A. Martel, G. London, and T. Alves, *Valuing form and function: Perspectives from practitioners about the costs and benefits of good apartment design.* In: *7th International Urban Design Conference*, 2014. Adelaide.

4. OECD, *HM1.5. Residential stock by dwelling type.* 2021, OECD. https://www.oecd.org/social/family/HM1-5-Housing-stock-by-dwelling-type.pdf

5. Moore, T., and A. Doyon, *The uncommon nightingale: Sustainable housing innovation in Australia.* Sustainability, 2018. **10**(10): p. 3469.

6. Horne, R., *Housing sustainability in low carbon cities.* 2018, London: Taylor & Francis Ltd.

7. Kelly, J.-F., *The housing we'd choose.* 2011, Melbourne: Grattan Institute. https://grattan.edu.au/wp-content/uploads/2014/04/090_cities_report_housing_market.pdf

8. Mann, B. *Surfside Officials: We weren't notified of severe deterioration before condo collapse.* 2021, 11/19/2021 https://www.npr.org/sections/live-updates-miami-area-condo-collapse/2021/07/08/1013457966/officials-not-notified-of-severe-deterioration-before-surfside-condo-collapse

9. United States Census Bureau, *2019 National – General Housing Data – All occupied units. Variable 1: Units by structure type.* 2019, United States Census Bureau. https://www.census.gov/programs-surveys/ahs/data/interactive/ahstablecreator.html?s_areas=00000&s_year=2019&s_tablename=TABLE1&s_bygroup1=3&s_bygroup2=1&s_filtergroup1=1&s_filtergroup2=1

10. O'Neill, J.P., *Bulletin 2. Summary of dwellings. Part 9. Australia.* 1971, Canberra: Commonwealth Bureau of Census and Statistics.

11. Australian Bureau of Statistics, *2071.0 – Census of Population and Housing: Reflecting Australia – Stories from the Census, 2016.* 2017, Australian Bureau of Statistics, 10/26/2021 https://www.abs.gov.au/ausstats/abs@.nsf/Lookup/by%20Subject/2071.0~2016~Main%20Features~Apartment%20Living~20

12. Treffers, S., *The emerging architecture of state regulation in North American condominium governance.* In: *Condominium governance and law in global urban context.* 2021, Routledge. pp. 197–216.

13. Easthope, H., *The politics and practices of apartment living.* 2019, Edward Elgar Publishing.

14. Lippert, R.K., and S. Treffers, *Introduction: Condominium governance and law in global urban context.* In: *Condominium governance and law in global urban context.* 2021, Routledge. pp. 1–9.
15. Raff, S., *The body corporate handbook: A guide to buying, owning and living in a strata scheme or owners corporation in Australia.* 2010, John Wiley & Sons.
16. Mandič, S., and M. Filipovič Hrast, *Homeownership in multi-apartment buildings: Control beyond property rights.* Housing, Theory and Society, 2019. **36**(4): pp. 401–425. DOI: 10.1080/14036096.2018.1510853.
17. Easthope, H., and A. Tice, *Children in apartments: Implications for the compact city.* Urban Policy and Research, 2011. **29**(4): pp. 415–434. DOI: 10.1080/08111146.2011.627834.
18. Gower, A., *Energy justice in apartment buildings and the spatial scale of energy sustainable design regulations in Australia and the UK.* Frontiers in Sustainable Cities, 2021. **3**(27). DOI: 10.3389/frsc.2021.644418.
19. Foster, S., P. Hooper, A. Kleeman, E. Martino, and B. Giles-Corti, *The high life: A policy audit of apartment design guidelines and their potential to promote residents' health and wellbeing.* Cities, 2020. **96**: p. 102420. DOI: 10.1016/j.cities.2019.102420.
20. Willand, N., and M. Nethercote, *Smoking in apartment buildings – Spatiality, meanings and understandings.* Health & Place, 2020. **61**: p. 102269. DOI: 10.1016/j.healthplace.2019.102269.
21. Oswald, D., T. Moore, and E. Baker, *Post pandemic landlord-renter relationships in Australia,* AHURI Final Report No. 344. 2020, Melbourne: Australian Housing and Urban Research Institute Limited. https://www.ahuri.edu.au/research/final-reports/344; DOI: 10.18408/ahuri5325901.
22. Wodnicki, J. *Condo board letter.* 2021. https://apps.npr.org/documents/document.html?id=20974547-condo-board-l
23. Moore, T., R. Horne, and A. Doyon, *Housing industry transitions: An urban living lab in Melbourne, Australia.* Urban Policy and Research, 2020. pp. 1–14 DOI: 10.1080/08111146.2020.1730786.
24. Dupuis, A., S. Blandy, and J. Dixon, *Introduction.* In: A. Dupuis, S. Blandy, and J. Dixon, eds. *Multi-owned housing: Law, power and practice,* 2010. Milton: Taylor & Francis Group.
25. Easthope, H., B. Randolph, and S. Judd, *Managing major repairs in residential strata developments in New South Wales.* 2009, Sydney: City Futures Research Centre, University of New South Wales.
26. McKenzie, E., *Emerging regulatory trends, power and completing interests in US common interest housing developments.* In: A. Dupuis, S. Blandy, and J. Dixon, eds. *Multi-owned housing: Law, power and practice,* 2010. Milton: Taylor & Francis Group.
27. Robertson, D., *Disinterested developers, empowered managers and vulnerable owners: Power relations in multi-occupied private housing in Scotland.* In: A. Dupuis, S. Blandy, and J. Dixon, eds. *Multi-owned housing: Law, power and practice,* 2010. Milton: Taylor & Francis Group.
28. van der Merwe, C., *European and South African law perspectives on the efficacy of sanctions to confront chronic rulebreakers in condominium developments.* In: *Condominium governance and law in global urban context,* 2021. Routledge. pp. 130–143.
29. Moore, T., N. Willand, S. Holdsworth, S. Berry, D. Whaley, G. Sheriff, A. Ambrose, and L. Dixon, *Evaluating the cape: Pre and post occupancy evaluation update January 2020.* 2020, Melbourne: RMIT University and Renew. https://renew.org.au/wp-content/uploads/2020/01/Evaluating-The-Cape-research-RMIT_Renew-January-2020.pdf

30. Byrne, J., S. Berry, and C. Eon, *Transitioning to net zero energy homes—Learnings from the CRC's high-performance housing living laboratories*. In: P. Newton, et al., eds. *Decarbonising the built environment: Charting the transition*, 2019. Singapore: Palgrave Macmillan. pp. 143–162.

31. Berry, S., and K. Davidson, *Value proposition: Low carbon housing policy*. 2015, Adelaide: University of South Australia. http://www.unisa.edu.au/Global/ITEE/BHI/Lochiel%20Park/BerryDavidson_ValueProposition-GovernmentExperience3.pdf

32. Pears, A., and T. Moore, *Decarbonising household energy use: The smart meter revolution and beyond*. In: *Decarbonising the built environment*, 2019. Springer. pp. 99–115.

33. Moore, T., F. de Haan, R. Horne, and B. Gleeson, *Urban sustainability transitions*. *Australian cases – International perspectives*. Theory and practice of urban sustainability transitions. 2018, Singapore: Springer.

34. Moore, T., I. Ridley, Y. Strengers, C. Maller, and R. Horne, *Dwelling performance and adaptive summer comfort in low-income Australian households*. Building Research & Information, 2017. pp. 1–14. DOI: 10.1080/09613218.2016.1139906.

35. Felder, M., *Strong, weak and invisible ties: A relational perspective on urban coexistence*. Sociology, 2020. **54**(4): pp. 675–692. DOI: 10.1177/0038038519895938.

36. Hirvonen, J., and J. Lilius, *Do neighbour relationships still matter?* Journal of Housing and the Built Environment, 2019. **34**(4): pp. 1023–1041. DOI: 10.1007/s10901-019-09656-0.

37. Premium Strata. *Strata manager, building manager, property manager. What is the difference?* 2021, 11/25/2021 https://premiumstrata.com.au/common-questions/strata-manager-building-manager-property-manager-difference/

38. Levy, D., H.C. Perkins, and D. Ge, *Improving the management of common property in multi-owned residential buildings: Lessons from Auckland, New Zealand*. Housing Studies, 2019. pp. 1–25. DOI: 10.1080/02673037.2018.1563672.

39. Easthope, H., and B. Randolph, *Governing the compact city: The challenges of apartment living in Sydney, Australia*. Housing Studies, 2009. **24**(2): pp. 243–259. DOI: 10.1080/02673030802705433.

40. Bounds, M., *Governance and residential satisfaction in multi-owned developments in Sydney*. In: A. Dupuis, S. Blandy, and J. Dixon, eds. Multi-owned housing: Law, power and practice, 2010. Milton: Taylor & Francis Group.

41. D'Souza, T., *The story behind a downtown Sarasota condo's near-collapse*. 2021, *Sarasota Magazine*, 10/26/2021 https://www.sarasotamagazine.com/home-and-real-estate/an-unlikely-heroine-steps-in-to-save-crumbling-dolphin-tower

42. Moore, T., and D. Oswald, *Why did the Miami apartment building collapse? And are others in danger?* 2021, The Conversation. https://theconversation.com/why-did-the-miami-apartment-building-collapse-and-are-others-in-danger-163425

43. Anonymous, *Email related 2018 structural field survey report*. 2018. https://www.townofsurfsidefl.gov/docs/default-source/default-document-library/town-clerk-documents/champlain-towers-south-public-records/email-records/email-related-2018-structural-field-survey-report.pdf?sfvrsn=aa311194_2

44. Baker, M., A. Singhvi, and P. Mazzei, *Engineer warned of 'major structural damage' at Florida Condo Complex*. 2021. https://www.nytimes.com/2021/06/26/us/miami-building-collapse-investigation.html

45. Lewis, R. *Months before Florida Condo collapsed, residents and the board sparred over repairs*. 2021, 11/19/2021 https://www.npr.org/sections/live-updates-miami-area-condo-collapse/2021/07/02/1012373938/months-before-florida-condo-collapse-residents-and-board-sparred-over-repairs

46. Anderson, C., and B. Condon, *Report showed 'major' damage before Florida Condo collapse.* 2021. https://apnews.com/article/fl-state-wire-florida-2a241993956ea842262e593812ad3ada
47. Oswald, D., *Homeowner vulnerability in residential buildings with flammable cladding.* Safety Science, 2021. **136**. DOI: 10.1016/j.ssci.2021.105185.
48. Oswald, D., T. Moore, and S. Lockrey, *Flammable cladding and the effects on homeowner well-being.* Housing Studies, 2021. pp. 1–20. DOI: 10.1080/02673037.2021.1887458.
49. Oswald, D., T. Moore, and S. Lockrey, *Combustible costs! Financial implications of flammable cladding for homeowners.* International Journal of Housing Policy, 2021. pp. 1–21. DOI: 10.1080/19491247.2021.1893119.

7 Navigating landlord-tenant conflicts

Key takeaways

- The housing consumer is not restricted to owner-occupiers but also landlords and tenants. This can create extra complexity when cracks, cladding, crisis, or other challenges emerge.
- Private rental housing is often amongst the poorest quality housing stock, meaning there are more likely to be quality, performance, and cost challenges for landlords and tenants to negotiate to deliver improved social value.
- The COVID-19 crisis has amplified problems within the private rental sector in many regions of the world, including tenants reluctance to report defects due to concerns of housing insecurity, and refusal and delays for tenants to have defects fixed. This has negatively impacted on tenant social value.
- Both landlords and tenants as housing consumers have been impacted financially by the COVID-19 crisis leaving greater challenges to delivering social value.
- The chapter presents a framework for landlords and tenants to negotiate rental payment challenges for those affected. This framework could be expanded to help resolve housing quality and performance issues, such as defects, and help improve and deliver wider social value for landlords and tenants.

Chapter summary

While home ownership remains the dominant housing tenure in many regions of the world, private rental housing is one of the fastest growing forms of housing tenure. Researchers around the world have found that the housing quality and performance in the rental sector is often amongst the poorest quality housing stock and that often those who are most socially and financially vulnerable in societies are those who live in such housing. In this chapter we explore the emerging evidence around the landlord and tenant relationship and what this means in the context of addressing defects. Our own research is presented of tenants and landlords during the early months of the COVID-19 pandemic and we discuss the key implications for social value (including well-being, finance,

DOI: 10.1201/9781003176336-7

and addressing defects). A number of key lessons are presented, with a potential framework to address not just rent reduction during COVID-19 but that could also be used to help landlords and tenants negotiate on minor and major defect repairs and upgrades.

7.1 An introduction to private rental housing

The consumer of residential construction is often thought of as the owner-occupier of a residential dwelling. However, there is also a significant proportion of consumers that own a dwelling and rent it out to a tenant. This is becoming more common as housing becomes increasingly unaffordable for a growing percentage of the population, more people are prioritising flexibility and location over home ownership and an investment in property can be seen as a wealth generation mechanism [1–7].

The percentage of private rental housing across the world varies. Across the OECD countries the private rental rate is about 23% [8]. However, in owner-occupier-dominated countries like Romania and Croatia the private rental housing stock is less than 2% of their total housing. Around a third of the total housing stock is private rental housing in countries like the United Kingdom, United States, and Australia and around half of all housing is private rental in Switzerland, Germany, and Denmark [8].

While previous chapters in this book have explored the consumer typically through the lens of an owner-occupier, it is important to recognise that there can be a landlord-tenant dynamic that adds complexity to emergent consumer issues presented throughout this book. The importance of this dynamic has become clear as a range of issues relating to building quality, performance, and safety have increasingly impacted tenants (as occupants) and landlords (as owners) in recent decades. The fundamental problems with building quality and creating social value in the private housing sector (that have emerged in recent decades) have been exacerbated through crisis situations like COVID-19 or responding to the flammable cladding crisis (see Case Study 7.1) [9–14].

Case Study 7.1   Leaseholders in the United Kingdom cladding crisis.

In England and Wales, the leasehold system exists. It is a long tenancy, where when someone buys a property, it is for a certain period of time (e.g. 99 years) [15]. At the end of this period, the lease ends, and the property is returned to the landlord [15]. Within the leasehold system, the landlord nominally owns the building and land, while the leaseholders pay all the bills including ground rents, insurance, service fees, maintenance, and a range of other costs [15]. The latest bill being pushed onto leaseholders is the rectification costs of flammable cladding.

The costs of flammable cladding being left for leaseholders have caused anger, protest, and debate in government. Within government, an amendment to stop landlords passing on flammable cladding costs to leaseholders has been voted against several times [16, 17]. While some financial government assistance is available, there are many leaseholders who are being left to pay the costs for flammable cladding rectification. This is particularly the case for buildings under 18 m tall (approximately 6 storeys), where limited government support is available (to date) [18]. Research into the effects on leaseholders has found that it has had a catastrophic impact on the mental health of some leaseholders, with a loss of control over their won lives, an inability to plan, no timeframes for rectification, and crippling financial burdens [19]. The case against the landlord pushing financial costs onto leaseholders continues.

This cladding crisis in the United Kingdom has all manifested since the Grenfell disaster in 2017. In the unfolding Grenfell inquiry, it is becoming clear that there were issues with landlords prior to the Grenfell Disaster occurring. For example, in 2016 (prior to the Grenfell Disaster), the Grenfell Action Group stated:

> It is a truly terrifying thought but the Grenfell Action Group firmly believe that only a catastrophic event will expose the ineptitude and incompetence of our landlord … and bring an end to the dangerous living conditions and neglect of health and safety legislation that they inflict upon their tenants and leaseholders … we will do everything in our power to ensure that those in authority know how long and how appallingly our landlord has ignored their responsibility to ensure the health and safety of their tenants and leaseholders. They can't say that they haven't been warned! [20].

The leasehold system demonstrates the complexity of different types of consumers, including the landlord and leaseholder. The outcomes and implications of the cladding crisis for these consumers are significant. In terms of financial costs, well-being of residents, and the lack of social value placed upon many on these homes with cladding. Many homes with flammable cladding currently have no financial worth and are a huge liability rather than an asset to be proud of.

Rental housing (including private rental, but also other types such as public or social housing) is important for a range of reasons, such as ensuring everyone has access to safe and secure housing. In countries like Australia, private rental housing stock has long been seen as the stepping stone for people between living in the family home and entering housing ownership on their own. While the numbers of private (and public and social) rental housing have fluctuated across time, for a range of factors (including affordability, ability to not be fixed to one location) many countries are seeing an increasing percentage of private rental housing stock and private rental tenants in recent years [8].

While there are some larger-scale private rental landlords, in many countries a significant percentage of private landlords are 'mum and dad' landlords who own a small number of investment properties. In the United States, for example, landlords own on average two dwellings [21]. In New Zealand, around 80% of landlords only own one investment property [22]. In Australia about 20% of households are residential property investors, with 71% owning one rental property, another 19% owning two properties, and less than 1% owning 6 or more [23]. In the United Kingdom, 45% of landlords own one property, with another 38% owning between 2 and 4 properties with an average of 1.9 properties per landlord [24]. This is important to point out as it highlights that in many countries a significant portion of rental housing is owned by people who are unlikely to have substantial experience in managing properties.

Wealth creation through property ownership has long been seen as key to building individual wealth. This has increasingly extended from owning your own dwelling to building wealth through investment properties. Governments like private rental housing because it helps provide housing outside of the public or social rental housing stock to ensure sufficient housing is provided for those who are unable, or chose not to, buy their dwelling. In many countries, there is a range of financial and tax benefits in place to encourage more people to become landlords and purchase investment properties. For example, negative gearing tax incentives in Australia essentially allow landlords to claim a loss of income from the rental property against income tax, helping to reduce taxable income and provide better tax outcomes for investors. Such financial or tax incentives lead to issues such as impacting on housing affordability but also contribute to the ongoing perception that landlords must be wealthy. However, many are small-scale investors, with research revealing it is not just the highest income earners who are landlords [23, 24]. These small-scale 'mum and dad' investors are often facing their own financial challenges to service mortgages with unexpected financial surprises that can tip them into financial stress. These unexpected financial surprises range in scale and cost, such a leaky tap and broken appliances, through to larger-scale issues that have emerged such as COVID-19 or rectification work for flammable cladding.

In the early months of the COVID-19 pandemic it was reported that in countries like Australia, half of the rental providers were experiencing mortgage stress and many were at risk of defaulting on housing repayments [25]. The income loss experienced both by renters and landlords as a result of the pandemic has caused high levels of financial uncertainty, emotional stress, and anxiety [10, 11, 13, 14, 26–34]. While various government and financial institution support such as pausing mortgage repayments helped to address these concerns for a period of time, it mostly pushed the issue down the road.

The ban on evictions in countries like the United States, United Kingdom, and Australia was critical for keeping tenants in a home during the pandemic. However, the support provided to allow tenants to pay reduced rent and/or defer payments until a later stage caused new issues for tenants once the ban on evictions was lifted. For example, in the United Kingdom, the lifting of the eviction ban led to hundreds of tenants who were in rent arrears being evicted, with one report finding that evicted tenants were on average £6500 in rent arrears [35]. The issue is compounded for many evicted households who have lost financial security during the pandemic (both through loss of jobs and having to use any savings to pay for rent, etc.). They have been left to find new housing with limited financial resources.

The social value challenges for consumers within private rental housing have escalated due to housing system weaknesses that have been further exposed by the COVID-19 crisis [10–13, 25, 27, 28, 34, 36–38].

7.2 Rental housing challenges: Landlord-tenant relationships, building quality, and creating social value

Rental housing is different from owner-occupied housing in numerous ways. There is evidence that around the world there are challenges with the relationships between landlords and renters [39–44]. The relationship is impacted by the real, and perceived, power dynamics which are re-enforced by governance structures and regulations that are often heavily in favour of protecting landlords. This frequently leads to tensions between landlords and tenants, which can significantly affect tenants' quality of life [45]. For example, tensions can arise due to the uncertainty over the security of the rental arrangements, the limited alternative housing options for tenants, and the lack of regulatory protection for tenants [46, 47].

Furthermore, tenants are often disadvantaged by the rental housing stock being over-represented by poor dwelling quality and performance in jurisdictions like the United States, United Kingdom, and Australia [39, 48, 49]. The quality and performance of a dwelling has a significant impact on the economic and social well-being of its occupants [50–60]. When combined with issues around stability and security of housing, this further negatively impacts tenant social and economic well-being [46, 48–50, 54, 59–64]. The recent work by Chisholm et al. [39] highlights the problem of private tenants living in cold, damp, or poorly maintained and unaffordable housing, despite the existence of housing codes and protective regulation.

Recent research into the private rental market in Australia found that 35% of private rental tenants reported that their rental dwelling required internal maintenance, 27% needed external maintenance, 30% were not adequately secure (e.g. good locks on doors and windows), 21% were affected by mould and damp and 14% were infested with pests (such as ants or cockroaches) [65]. Similar issues relating to the quality and performance of private rental quality have been found elsewhere. For example, in the United Kingdom, the English Housing Survey estimated that 23% of private rental housing did not meet the Decent Home Standard in 2019 and that they were more likely to have at least one Category 1 hazard under the Housing Health and Safety Rating System [66]. Another report in the United Kingdom found that across the previous two years 60% of private rental tenants surveyed had experienced disrepair that their landlord was responsible for fixing. This included 15% who said the repair was a major threat to their health and safety, almost 50% said it had a major impact on their level of comfort and almost 20% reported the repairs did not occur within a reasonable amount of time, with 10% stating that the repair was never completely fixed with some reporting having to spend their own money on the issue [67]!

One of the challenges with the private rental space is that there are two key actors involved that could be considered consumers: the landlord and the tenant. The landlord is the person who owns the property and therefore must adhere to certain property ownership rights, but they have the ultimate power over the dwelling. We can think of the landlord as a consumer as they are the legal owner/customer of the property. A tenant, on the other hand, sits in a different position as someone who pays rent to use and live in the home. Thus, they have entered into a legal agreement with the landlord to compensate them a set amount to be able to live in the landlord's property for a defined period of time. Depending on where you are in the world, there will then be a set rules about what a tenant can do within, and to, a rented property. For example, in some jurisdictions a tenant may not be able to hang a picture with a permanent fixture (e.g. a nail) without getting landlord permission, where, in other jurisdictions, as long as the property is at least left in the same (or better) condition than when the tenants moved in, then the tenants can do a range of things (including smaller improvements) to the property. In this way, tenants are (housing) consumers but are also end-users as they must adhere to the requirements out by the landlord and other legal structures.

There is also a range of other laws and regulations in each jurisdiction which attempt to set clear frameworks for the landlord and tenant relationship and to provide levels of protection for both parties. The legal requirements differ across jurisdictions but typically cover issues around the setting and collection of rent, tenant rights (e.g. rules around eviction) and the process for dealing with disputes. These requirements attempt to address the imbalance of power in the landlord-tenant relationship.

In more recent years, these legal requirements have extended in some locations, such as the United Kingdom and Europe, to the setting of minimum quality of performance standards in rental housing. These typically cover things that relate to the basic functioning of the dwelling, such as making sure there are no leaking taps, there is a working heating and cooling system, and any provided appliances work. For example, since April 2020 in the United Kingdom private landlords must comply with the Domestic Minimum Energy Efficiency Standard Regulations [68]. For any property which has an Energy Performance Certificate rating of F or G (where A is the best rating) the energy performance must be improved to at least a level E. The Regulations do have some exceptions and a landlord is only obliged to spend up to £3500 to undertake the improvements. In New Zealand, the Residential Tenancies (Healthy Homes Standards) Regulations 2019 has established minimum quality and performance requirements of rental dwellings as outlined in Table 7.1 [69].

The above examples of minimum standards are about ensuring that private rental housing meets certain quality and performance at the point it is leased. However, once a tenant moves into a private rental dwelling, then the responsibility for maintenance and upkeep of the property depends on the jurisdiction and legal settings. In many other countries, such as Australia, the private

Table 7.1 New Zealand Residential Tenancies (Healthy Homes Standards) Regulations 2019 overview [69].

Standard	Required standard
Heating	There must be fixed heating devices, capable of achieving a minimum temperature of at least 18°C in the living room only. Some heating devices are inefficient, unaffordable, or unhealthy and will not meet the requirements under the heating standard.
Insulation	The minimum level of ceiling and underfloor insulation must either meet the 2008 Building Code or (for existing ceiling insulation) have a minimum thickness of 120 mm.
Ventilation	Ventilation must include openable windows in the living room, dining room, kitchen, and bedrooms. Also an appropriately sized extractor fan(s) in rooms with a bath or shower or indoor cooktop.
Moisture ingress and drainage	Landlords must ensure efficient drainage and guttering, downpipes, and drains. If a rental property has an enclosed subfloor, it must have a ground moisture barrier if it is possible to install one.
Draught stopping	Landlords must stop any unnecessary gaps or holes in walls, ceilings, windows, floors, and doors that cause noticeable draughts. All unused chimneys and fireplaces must be blocked.

landlord is responsible for the maintenance and upkeep of the property. In these situations, there is still an expectation that tenants look after the property but that they are protected if normal wear and tear on the property occurs. If there is significant damage due to the tenants' actions, the tenant would typically be liable for fixing the issue. Table 7.2 shows tenant rights and responsibilities in the United Kingdom [70].

Table 7.2 Tenant rights and responsibilities in the United Kingdom [70].

Tenant rights	Tenant responsibilities
• Live in a property that is safe and in a good state of repair • Have deposit returned when the tenancy ends • Challenge excessively high charges • Know who your landlord is • Live in the property undisturbed • See an Energy Performance Certificate for the property • Be protected from unfair eviction and unfair rent • Have a written agreement if you have a fixed-term tenancy of more than three years • If you have a tenancy agreement, it should be fair and comply with the law	• Give landlord access to the property to inspect it or carry out repairs. Landlord has to give at least 24 hours' notice and visit at a reasonable time of day, unless it is an emergency and they need immediate access • Take good care of the property, for example, turn off the water at the mains if you are away in cold weather • Pay the agreed rent, even if repairs are needed or you are in dispute with your landlord • Pay other charges as agreed with the landlord, for example, Council Tax, or utility bills • Repair or pay for any damage caused by you, your family, or friends • Only sublet a property if the tenancy agreement or your landlord allows it

The various legal requirements in many jurisdictions are also meant to cover issues, such as defects. This is through setting out clear responsibilities for landlords or tenants to address any issues that arise with the basic quality and performance of a rental dwelling. In many countries, the onus is typically on the tenant to report any housing quality or performance issues. Research has found that there is a significant issue around the power imbalance towards landlords, meaning many tenants do not ask for defects to be fixed, or improved. This is due to concerns that they could be evicted, have their rent increased, or not receive a favourable reference from the landlord when they rent their next property [71–73]. Another issue that tenants face is that in many countries it is common for landlords to be represented by a professional agency acting on their behalf (such as a real estate agent), whereas tenants generally have to seek out any professional assistance on their own and often lack clear information about their rights [74, 75]. This also means many minor and larger property defects go unreported (see Case Study 7.2).

Case Study 7.2   The process for defect resolution in private rental housing in Victoria, Australia.

In the State of Victoria, Australia, a process for having defects and other tenant-landlord-related issues addressed has been developed by Consumer Affairs Victoria [76]. This is an attempt to make it clear what the steps are and what the rights are for tenants and what responsibilities are for landlords. Broadly, the process is that once a tenant is aware of a defect of maintenance issue, they need to inform the rental provider (landlord or real estate agent if one is used) by phone and it is recommended they follow up this with a written request to act as evidence of the request. As long as the issue is not the renter's fault then the landlord must pay for all the repairs. Urgent repairs must be undertaken immediately, with non-urgent repairs to be undertaken within 14 days of getting a written repair request. However, evidence shows that this process is not always translated into practice.

Urgent repairs are defined by law and include things such as:

- burst water service;
- blocked or broken toilet system;
- serious roof leak;
- gas leak;
- dangerous electrical fault;
- flooding or serious flood damage;
- serious storm or fire damage;
- an essential service or appliance for hot water, water, cooking, heating, or laundering is not working;
- the gas, electricity or water supply is not working;
- a cooling appliance or service provided by the rental provider is not working;
- the property does not meet minimum standards;
- a safety-related device, such as a smoke alarm or pool fence, is not working;

- an appliance, fitting or fixture that is not working and causes a lot of water to be wasted; and
- any fault or damage in the property that makes it unsafe or insecure, including pests, mould, or damp caused by or related to the building structure.
- a serious problem with a lift or staircase.

If the rental provider does not respond in a timely manner to an urgent repair request, the renter can organise and pay for the repair up to the cost of AU$2500. This cost must be repaid to them by the rental provider within seven days. If a tenant is unable to afford this cost, then they can contact the Victorian State Government for assistance.

If an urgent or non-urgent repair is not undertaken within the set time periods, or there are issues with the repair work undertaken, or the tenant has not been repaid for an urgent repair, there is a process the tenant can follow to have the issue resolved, including lodging a complaint via the Victorian Civil and Administrative Tribunal [77]. This Tribunal will hear urgent repair claims within two business days and non-urgent claims within seven days. The Tribunal will hear all the evidence about the issue and deliver an order which says what the findings and outcomes are. The findings are legally enforceable.

Despite the above process in Victoria (which is similar in other states in Australia), tenants are often left waiting for a significant period of time to have even the most minor of issues addressed. Research from Australia has found that up to 60% of private rental tenants had to wait more than a week to have a maintenance request addressed, and 18% said it was only fixed with constant reminders to the landlord or real estate agent [65]. Of concern is that just under 10% of tenants said they either gave up on getting the issue fixed due to a lack of response and decided to just live with the issue or fix it themselves. This process also puts much of the responsibility for driving the process along on the tenant. Given that tenants can often feel insecure in their housing and feel they have a lack of rights and protection, this likely means that many defect issues go unreported or are not pursued if the tenant comes up against resistance. While there is a process in place, it needs to be evolved to provide better protection and clarity for tenants.

7.3 Fixing defects during a crisis

Tenants face a challenge of relying on the landlord to address significant quality and performance property issues which can have implications on social value. The landlord, as a consumer that is not living in the property, may be reluctant to spend money on defects or improve the quality of the dwelling if they do not see it result in increased financial returns. This potential tension between tenants and landlords has been exacerbated by the COVID-19 crisis. The pandemic has changed the function of the home to also be a public health intervention, since households have needed to rely on their dwellings as part of the protection mechanisms against the virus [9, 33, 38, 78]. Across the world, isolating at home and undertaking activities like work or education from home during the pandemic has resulted in a change to social value in the home and highlighted issues of housing security, quality, and conditions, especially in the private rental sector [33, 38, 78]. It was widely reported, during the crisis, that tenants in different countries were facing additional difficulties in having basic maintenance and upkeep addressed, in addition to having defects fixed. This is despite many jurisdictions making it clear that landlord and tenant rights and responsibilities needed to be carried out even during these situations (as long as any repair work was able to be undertaken by safely adhering to any local COVID-19 restrictions).

Our research during the second wave of the COVID-19 pandemic in Australia highlights that there were challenges for some tenants to have defects addressed and that this could impact the useability and liveability of the dwelling significantly [9]. In survey and interview responses, some tenants felt that landlords and property agents were using the COVID-19 pandemic as an excuse not to undertake required repairs. As one tenant stated:

> There have been a lot of problems ... that the landlords haven't dealt with or fixed...Five power outlets do not work. All electronics through extension cables through the house, safety issues ... We have heard from them [the agent] that 'due to the COVID-19 pandemic, we are unable to get our tradesmen out there, and they are only responding to emergencies' ... ok whatever ... they are not emergencies, but they still need to be addressed.

Other tenants reported that minor defects had become more influential on their lives, as they spent more time at home. They explained they may not have reported these previously, since they had not significantly impacted their ability to live in the dwelling in the past. However, now with the extended periods of time at home, they were affecting their well-being. For example, one tenant reported that an issue of noisy rattling windows was impacting their ability to productively work from home:

> The windows make a lot of noise if it's raining or windy, and I have repeatedly emailed the landlord and real estate agent and haven't got anywhere ... they have had someone come and look at it, and they have said they need to replace the windows as they are that old, but the owners haven't and I don't think they are going to fork out the cash. I think why it has become a real issue is because I have now been working from home, and sometimes I have to wear earplugs, and on top of that, headphones, to block out the noise from the windows.

Another tenant was having a range of issues fixed and was disappointed with the lack of response from the property agent. Again, with the extra time being spent at home, it was compounding the issue and affecting the liveability of the house:

> The agent and the landlord have promised to fix the dishwasher and other issues, but I still haven't heard from them and do not expect to either ...

> I hate this place now, I hate this house, a lot of electrics plus don't work, the dishwasher doesn't work, the hot water is so limited, and very hard to control ....

There was also a concern from some tenants that if they asked for anything to be fixed during COVID-19 it might negatively impact the likelihood that the landlord would be open to discussing rent reductions or rent pauses due to the tenant losing their job or having reduced working hours due to the pandemic:

> We've had to raise some maintenance issues as we've had a leak and were worried about doing this while negotiating a reduced rent but the landlord approved our gutters to be cleaned immediately and some roof tiles replaced.

Another said that they requested something to be fixed but then decided to drop it as they were concerned about future rental references:

> We decided not to complain about it further, as they [landlord] might say we were not a good renter, so we decided not to take it any further.

However, this was not evident in all cases, with some tenants revealing that there had been no issues with some of the maintenance requests they had made. One tenant said:

> Some of the windows wouldn't close all the way, we asked someone to come and fix that and they sent someone right away to fix that.

The challenges with getting repair work undertaken in a timely manner were not just reported by tenants, with some landlords reporting their own property agents had been slow to organise repair work. One landlord reported:

> [I] have received no follow up on a maintenance issue from our agent. The issue was a minor repair to our deck at our investment property. The issue has been open for 3 months. Our assumption is the hold-up is due to COVID-19 and availability of tradespeople ... I mentioned to the agent that I was concerned it would leave our tenant feeling as though the issue had not been addressed.

In addition, the poor quality and performance of rental housing were creating unexpected financial challenges for tenants who were spending more time at home. In our research we found that tenants reflected that the poor quality of their dwellings, and the fact they were spending more time at home during lockdowns, meant their utility bills had gone up significantly, largely due to the high costs for heating and cooling thermally poor dwellings.

> I am waiting for the [energy] bill, it will be huge, I am scared ... but I also know we cannot avoid an increase in the bill, cause we need the heater, cause

> two of us are working from home, and it is really cold, and also I have an extra [computer] screen ….

Some tenants were even energy rationing and potentially putting their health and well-being at risk:

> We haven't used the heater so far; it is getting very cold.

Our research highlighted how the COVID-19 crisis was exacerbating current problems for consumers in the private housing sector, including affordability for heating and cooling and responses to fixing defects. While there is often a clear process for both tenants and landlords to follow to have repair work undertaken or to revise rental conditions, these processes often require some form of communication and negotiation between the two parties. This process of negotiation has previously been noted as providing additional stress on tenants who often feel the power imbalance (discussed in Section 7.2). Our research has revealed that there is a real tension and challenge for many tenants to approach their landlord, even for things that are clearly defined within their tenant rights (such as addressing defects in a timely manner).

While these issues had implications for building quality and social value, there were also concerns around tenant housing security that were exacerbated by the COVID-19 crisis. The pandemic meant there was a need for negotiations and dialogue between tenants and landlords, particularly around any financial problems that have emerged, as a result of loss of employment or underemployment.

Our study over the early months of the pandemic looking at tenants and landlords in Australia found that more than 60% of tenants had, or would likely have in the near future, issues paying their rent [9]. While the Federal and State governments in Australia had introduced a range of additional protection measures for tenants (e.g. a pause on evictions) there was still a requirement for tenants to negotiate changes with landlords. However, some tenants in our study said they had not reached out to their landlords about issues of paying their rent (due to job losses or reduced working hours due to COVID-19). This was, in part, because they were concerned the landlord would then stop responding to requests for any maintenance of defect work. Others, as noted in the previous section, were not reaching out to their landlords to make maintenance requests for the opposite reason: that they were worried it would impact their ability to negotiate a rent reduction, if needed. This likely comes back to tenants' understanding of their rights, with a survey in Australia finding that just under half of tenants have a good understanding of their rights [65]. This lack of understanding is likely to impact the perception of what tenants can, and cannot, ask for.

What was clear though from both the tenants and landlords in our study was that there was no typical way in which these approaches and negotiations would start, or the process in which they would occur, and that previous ways for tenants and landlords to connect and build a relationship were not adequate during a crisis. For example, some tenants said they reached out to their landlords, while others reported their landlords being proactive and reaching out to their tenants first. At times this was not an issue, but in some cases, it created additional challenges for tenants and landlords. One tenant reflected:

> My partner and I rent a house through a private landlord … They are lovely … I work part time in local government and my partner works full time … He was dropped down to 4 days in May but we have few expenses and no debt so we can easily meet our bills still … Our landlords got in touch when lockdown began to see if we were ok and to let them know if we needed to negotiate rent etc. We feel extremely fortunate.

Our analysis found that there were two clear groups of landlords: those who were willing to consider financial assistance for their tenants and those who were against this. Many landlords reflected that during the pandemic we all needed to be more accommodating and make sacrifices and that they had a role and responsibility for the well-being of their tenants. It was not only about helping tenants, but also a reflection that receiving some money as rent was better than letting the property sit there empty, without rent, if the tenant had to move out:

> We've had good renters. They have been there for five years and always paid on time … What happened when COVID hit was we got a call from our real estate agent saying 'look this is what they are requesting' it was a no brainer for my wife and I. We didn't bother negotiating. They were renting for $500 a week and asked for $350 per week for 3 months … Given the tenants have been good and looked after the property let's not do the payment plan and just go with the reduction because at the end of the day that human kindness comes in … someone has to cut them a break.

Other landlords who had resisted reducing rental payments seemed to come from concerns around their own financial position, such as the need to continue to service their own mortgages or for landlords who were retired. This has impacts on their living costs significantly, as one such landlord said:

> When the agent rang me and said can I reduce their rent by $100 a week I said absolutely I can't do that because that is my income. I am living very frugally because I don't want to have to start selling things to live off.

This study showed that negotiations lacked a set of protocols and understanding of acceptable practice, and there were high levels of uncertainty around what happens once initial (or continued) financial support ends. This impacted the social value of landlords and tenants negatively, with both being squeezed financially. It is possible that this will create issues for housing quality and performance longer term with landlords who were financially hit during COVID-19 potentially less inclined to undertake maintenance and repair work to investment properties. In some cases, landlords will be left with no choice but to sell their investment property.

7.4 Next steps: Resolving landlord-tenant tensions

During a crisis, such as the COVID-19 pandemic, housing systems can be put under significant stress. This is because there are additional challenges to navigate with loss of employment, unpaid rent, and housing security. With properties also being used as a public health intervention during lockdowns, the function of a home has changed. This means there can be building quality issues that become more problematic (and need to be addressed), as people need to live, work, and leisure at home for extended periods. These issues around financial stability, housing security, and quality, all have implications for the social value that can be placed on a residence, especially during a time of crisis.

The relationship between landlords and tenants can influence well-being, financial vulnerabilities, and the social value placed on the rental property. As discussed above, there are challenges with tenants having defects fixed, and this has become even more difficult for some tenants during the COVID-19 pandemic. The role of negotiations around defects, but also other issues such as rent reductions, has highlighted the real and perceived power imbalance between tenants and landlords.

As noted earlier, the quality and performance of private rental housing is often amongst the poorest quality housing. Given the well-established association between health and quality of housing, a fall in housing standards, or a delay in fixing defects, could also have implications for health conditions, with the relationship between housing and health vulnerability potentially further heightened during a pandemic [9, 10, 79]. The evidence shows that tenants are not always clear about their rights, and there is a perception that if they reach out to their landlord to fix defects (or other concerns), and that this will impact their rental arrangement or have implications for their future rental security.

One of the outcomes that emerge not only from our own research but from international examples during COVID-19 is that despite there is often a clear process put forward by governments for how to address repairs and defects in a rental dwelling (and other issues such as rent arrears), that these processes are not sufficient, especially during a crisis. From our own research there are some clear recommendations which governments should look into not only to implement or improve in addressing outcomes in a crisis like COVID-19 but also to improve the landlord and tenant relationship more broadly. Future policy to support landlord-tenant relationships should consider:

- *Tightening the existing rights and responsibilities* of tenants and landlords. While many jurisdictions have clear timeframes around how long it should take a defect to be addressed, there are rental households who are not having this work completed satisfactorily within the prescribed period. Tenants are unlikely to feel secure in taking this matter forward to the relevant authorities for fear of repercussions from the landlord. A centralised database where defect requests can be lodged, tracked, and followed up on may provide improved protection for tenants and make it clear that landlords must deliver on their responsibilities or risk fines or other sanctions.

- In addition to this, there should be a *protective negotiation framework* developed to structure negotiations between tenants and landlords. This would allow tenants to request issues be addressed, such as defects, without fear of poor tenant references or refusal by the landlord to agree to the request. This should be led by an independent assessor that evaluates the situation in terms of what the defect is, the implications of it, possible solutions or outcomes. This could act as a middle step before more formal legal proceedings are attempted.
- *Clear government advice for real estate agents* to provide clarity, guidance, and information on acceptable and unacceptable practices in resolving landlord-tenant negotiations. For example, whether it is acceptable to suggest tenants should attempt to access their super/pension funds to address unpaid rent or whether they should allow tenants to speak directly with the landlord. This advice should also provide transparency during the negotiation process, with some tenants believing their information or requests were not being passed on to the landlord.
- *Improved information and guidance in defining the end of support*, such as protocols for rent deferral repayment plans. Landlords did not see repayment deferral plans as an option that provided much clarity about how tenants would repay costs and what would happen if they could not. Tenants also did not want to carry the additional burden of owing large sums to their landlord and therefore tried to negotiate a better deal or consider other options. It is recommended that during periods of crisis (such as lockdowns in the COVID-19 pandemic), there should be negotiations for rent-reductions rather than rent-deferrals. Such clarity at the end of the support period would help relieve uncertainty and provide more clarity during negotiations.

The Victorian State Government (in Australia) introduced a framework for how to make a temporary rent reduction agreement in November 2020. This framework was based, in part, on some of the wider industry feedback and early research which had been undertaken to look at the issues for renters and landlords [9]. It is proposed that this framework could be broadened to have a formal structured option for tenants, when requesting rent reduction during times of crisis, or building defect rectifications that they believe should be fixed. When landlord and tenant relationships are strong, these issues can often be dealt with informally. However, if there is no existing relationship, the option of a more formal and structured request may be appealing to tenants. A summary of the proposed framework is provided in Table 7.3.

The below formal and structured negotiation framework is another option for tenants to reach a resolution with their landlords. Resolutions of key issues can have clear benefits to the social value placed on the rental property, as tenants can feel more secure with their housing situations or appreciate an improvement in building quality.

This type of framework shows what is possible but also that this should come from regulators to ensure clarity and consistency. The checks at various steps (such as lodging outcomes) also help strengthen the process and provide protection for

Table 7.3 A proposed structured negotiation framework.

Step	Summary of action
Step 1 – Tenant works out what they can rightfully request, such as: • A rent reduction, based on how much rent is affordable for them during a time of crisis. • A building defect rectification, based on their rights as a tenant.	During a time of crisis, tenants will be impacted in different ways, with some losing jobs, having reduced hours, or not being affected financially at all. For those that are significantly affected, it is important to consider the ability to pay rent that is based upon their own unique circumstances rather than a general blanket rule. The amount that is affordable to the tenant needs to be 'reasonable in the circumstances' and the tenant needs to think about their income, any savings they have, and any essential expenses they would need to cover. This is not about trying to find the lowest price the landlord will accept but to determine what is fair and reasonable. There are a number of online budget calculators to help determine living expenses and how long any savings would last. This step could also be relevant to other requests, such as defect rectifications, where the first consideration is for the tenant to determine what they can request based on their rights.
Step 2 – Tenant makes a request (e.g. for a rent reduction agreement or building defect rectification).	Once the tenant has determined that their request is reasonable and within their rights, they should contact their landlord or property manager. In the case of a rent reduction, this should happen as soon as possible, but not before the tenant has a clear idea of what they can realistically afford to pay (step 1). For building defects, the substandard issues should be communicated as soon as practical with reference to tenant rights and an explanation of how this is affecting their living. There should be a number of template letters or example information that could be used to approach the landlord.
Step 3 – Negotiate an agreement (e.g. on a rent reduction or to fix a building defect).	Landlords can accept the tenant's request, such as to fix a building defect or a reduced rent request. They can also begin negotiating for a reasonable alternative. For example, landlords should be clear on what rent reduction they can accept based upon their own unique circumstances. The preference should be for a rent reduction, not deferral if possible (which may push issues down the road). Landlords and tenants should negotiate in good faith and be understanding of the other party's situation and be as flexible as possible.
Step 4 – Put it in writing and register it with a consumer affairs government body.	Once a tenant request has been agreed with, for example, a building defect, or a rent reduction, it should then be documented centrally. Again, templates of agreements should be available to help streamline the process. These agreements can then be lodged with a consumer body within the government, so there is evidence of the outcome.
Step 5 – If you cannot agree on the tenants' request (e.g. with a rent reduction or to fix a building defect).	If both parties are unable to agree they can start a dispute resolution process through the consumer body within the government, who will then work with both parties to help them reach an outcome. If this process still does not lead to an agreed outcome a Chief Dispute Resolution Officer may make a binding dispute resolution order. This process should be time bound to ensure prompt responses for both tenants and landlords.

both tenants and landlords. It also provides the basis for what could be developed to clarify and strengthen the process for tenants and landlords to request, negotiate and undertake repairs or upgrades to the property. Any potential requests should be lodged to a third party so that if issues arise in the future there is evidence available to help sort through any conflict. The process should have clear time frames to ensure that tenants and landlords can get prompt outcomes.

7.5 Conclusions

Private rental housing is one of fastest growing forms of housing as cities grow upwards and the construction of apartment buildings increases. The consideration of the consumer within the residential sector cannot be assumed to be an owner-occupier, with landlord and tenant arrangements creating more complexity at the consumer end of the supply chain. The private housing sector has been known to have weaknesses, such as the power imbalance between landlords and tenants that can result in defects being left unresolved, fears of eviction, and poor future references. These weaknesses were exacerbated when exposed more clearly by the COVID-19 pandemic.

There is a housing crisis that threatens to follow, with some financially restricted landlords reluctant to resolve housing quality issues and some tenants unable to pay rent and unable to afford heating and cooling. These challenges mean there is likely to be less and less consumers that are living with housing security, in a place that enhances their well-being and social value. Our research during the height of the second wave of the COVID-19 pandemic in Australia highlighted these challenges and recommended a protective framework for landlords and tenants to resolve consumer issues. While this framework was intended for the immediate resolution of rental disputes from the economic impact of the COVID-19 pandemic, it could be used beyond both the pandemic and beyond only financial disputes but also building quality issues. This protective framework could help tenants to feel comfortable to raise building quality issues, such as defects, without fear of eviction, poor future references, or raised rent.

References

1. Haffner, M., and K. Hulse, *A fresh look at contemporary perspectives on urban housing affordability.* International Journal of Urban Sciences, 2021. **25**(sup 1): pp. 59–79. DOI: 10.1080/12265934.2019.1687320.
2. Anacker, K.B., *Introduction: Housing affordability and affordable housing.* International Journal of Housing Policy, 2019. **19**(1): pp. 1–16. DOI: 10.1080/ 19491247.2018.1560544.
3. Wetzstein, S., *The global urban housing affordability crisis.* Urban Studies, 2017. **54**(14): pp. 3159–3177. DOI: 10.1177/0042098017711649.
4. Gurran, N., and P. Phibbs, *'Boulevard of broken dreams': Planning, housing supply and affordability in urban Australia.* Built Environment, 2016. **42**(1): pp. 55–71. DOI: 10.2148/benv.42.1.55.

5. Wood, G., R. Ong, and M. Cigdem, *Housing affordability dynamics: New insights from the last decade.* 2014, Melbourne: Australian Housing and Urban Research Institute.

6. Czischke, D., and G. van Bortel, *An exploration of concepts and polices on 'affordable housing' in England, Italy, Poland and the Netherlands.* Journal of Housing and the Built Environment, 2018. pp. 1–21. https://link.springer.com/article/10.1007/s10901-018-9598-1

7. Clapham, D., *Remaking housing policy: An international study.* 2018, Routledge.

8. OECD, *Housing market.* 2021. https://www.oecd.org/housing/data/affordable-housing-database/housing-market.htm

9. Oswald, D., T. Moore, and E. Baker, *Post pandemic landlord-renter relationships in Australia, AHURI Final Report No. 344.* 2020, Melbourne: Australian Housing and Urban Research Institute Limited. https://www.ahuri.edu.au/research/final-reports/344. DOI: 10.18408/ahuri5325901.

10. Bower, M., C. Buckle, E. Rugel, A. Donohoe-Bales, L. McGrath, K. Gournay, E. Barrett, P. Phibbs, and M. Teesson, *'Trapped', 'anxious' and 'traumatised': COVID-19 intensified the impact of housing inequality on Australians' mental health.* International Journal of Housing Policy, 2021. pp. 1–32 DOI: 10.1080/19491247.2021.1940686.

11. Ambrose, A., W. Baker, G. Sherriff, and J. Chambers, *Cold comfort: COVID-19, lockdown and the coping strategies of fuel poor households.* Energy Reports, 2021. DOI: 10.1016/j.egyr.2021.08.175.

12. Gurney, C.M., *Dangerous liaisons? Applying the social harm perspective to the social inequality, housing and health trifecta during the Covid-19 pandemic.* International Journal of Housing Policy, 2021. pp. 1–28. DOI: 10.1080/19491247.2021.1971033.

13. Martin, C., A. Sisson, and S. Thompson, *Reluctant regulators? Rent regulation in Australia during the COVID-19 pandemic.* International Journal of Housing Policy, 2021. pp. 1–21. DOI: 10.1080/19491247.2021.1983246.

14. AHURI, *Unpacking the challenges in the rental market during COVID-19. What are the policy options and how is each state faring?* 2020, AHURI Brief, 06/07/2020 https://www.ahuri.edu.au/news-and-media/covid-19/unpacking-the-challenges-in-the-rental-market-during-covid-19

15. O'Kelly, S., *The leasehold system is a money-making racket. Reform is long overdue.* 2018, The Guardian, 11/23/2021 https://www.theguardian.com/commentisfree/2018/jul/20/leasehold-money-making-racket-reform

16. *Cladding: MPs fail in bid to change fire safety payment rules.* 2021, British Broadcasting Corporation, 11/23/2021 https://www.bbc.com/news/uk-politics-56174558

17. *Grenfell: Government defeated on fire safety costs bill.* 2021, British Broadcasting Corporation, 11/23/2021 https://www.bbc.com/news/uk-politics-56905882

18. *Cladding: Why is it unsafe and what money has the government promised?* 2021, British Broadcasting Corporation, 11/23/2021 https://www.bbc.com/news/explainers-56015129

19. Corker, S., and J. Stewart, *Cladding crisis: 'Some days I can't leave the house'.* 2021, British Broadcasting Corporation, 11/23/2021 https://www.bbc.com/news/business-59320814

20. KCTMO, *Playing with fire!.* 2016, Grenfell Action Group Word Press, 11/23/2021 https://assets.grenfelltowerinquiry.org.uk/TMO00835660_GAG%20blog%20post%20-%20KCTMO%20Playing%20with%20fire.pdf

21. Richardson, T., *Message from PD&R Senior Leadership – Landlords.* 2018, The Department of Housing and Urban Development, 10/03/2021 https://www.huduser.gov/portal/pdredge/pdr-edge-frm-asst-sec-061118.html

22. Bell, M., *Nearly 80 per cent of landlords own just one property, data shows.* 2021, Stuff, 10/03/2021 https://www.stuff.co.nz/life-style/homed/real-estate/124320645/nearly-80-per-cent-of-landlords-own-just-one-property-data-shows

23. Australian Government, *Taxation statistics 2017–18.* 2020, Australian Taxation Office, 10/03/2021 https://data.gov.au/data/dataset/taxation-statistics-2017-18

24. Ministry of Housing, Communities and Local Government, *English private landlord survey 2018 main report.* 2019, London: UK Government.

25. AHURI, *COVID-19 mortgage stress creating uncertain housing futures. Pandemic recession leaves households in mortgage stress and socio-economic hardship.* 2020. https://www.ahuri.edu.au/policy/ahuri-briefs/covid-19-mortgage-stress-creating-uncertain-housing-futures

26. Casey, S., and L. Ralston. *Coronavirus puts casual workers at risk of homelessness unless they get more support.* 2020, 04/30/2020 https://theconversation.com/coronavirus-puts-casual-workers-at-risk-of-homelessness-unless-they-get-more-support-133782

27. Jones, A., and D.S. Grigsby-Toussaint, *Housing stability and the residential context of the COVID-19 pandemic.* Cities & Health, 2020. pp. 1–3. DOI: 10.1080/23748834.2020.1785164.

28. Maalsen, S., D. Rogers, and L.P. Ross, *Rent and crisis: Old housing problems require a new state of exception in Australia.* Dialogues in Human Geography, 2020. **10**(2). DOI: 10.1177/2043820620933849.

29. Rosenberg, A., D.E. Keene, P. Schlesinger, A.K. Groves, and K.M. Blankenship, *COVID-19 and hidden housing vulnerabilities: Implications for health equity, new haven, Connecticut.* AIDS and Behavior, 2020. **24**(7): pp. 2007–2008. DOI: 10.1007/s10461-020-02921-2.

30. Christensen, S., *Rebalancing the landlord and tenant relationship during the COVID emergency – State and territory adoption of the national leasing code.* Australian Property Law Bulletin, 2020. **35**(4): pp. 50–51.

31. Manville, M., P. Monkkonen, and M. Lens, *COVID-19 and renter distress: Evidence from Los Angeles. August 2020.* 2020, UCLA. https://escholarship.org/content/qt7sv4n7pr/qt7sv4n7pr.pdf

32. Phillips, S., *LA's COVID-19 response should prioritize long-term rent-stabilized tenants for housing assistance. May 2020.* 2020, UCLA. https://escholarship.org/content/qt1zv0m2gn/qt1zv0m2gn.pdf

33. Farha, L., *COVID-19 guidance note: Protecting renters and mortgage payers.* 2020, United Nations. 07/08/2020 https://www.ohchr.org/Documents/Issues/Housing/SR_housing_COVID-19_guidance_rent_and_mortgage_payers.pdf

34. Horne, R., N. Willand, L. Dorignon, and B. Middha, *Housing inequalities and resilience: The lived experience of COVID-19.* International Journal of Housing Policy, 2021. pp. 1–25. DOI: 10.1080/19491247.2021.2002659.

35. Buchanan, M., *Covid debt: A baby, job loss – and now eviction.* 2021, BBC News, 10/03/2021 https://www.bbc.com/news/uk-58643437

36. Bentley, R., and E. Baker, *Housing at the frontline of the COVID-19 challenge: A commentary on 'rising home values and Covid-19 case rates in Massachusetts'.* Social Science & Medicine, 2020. **265**: p. 113534. DOI: 10.1016/j.socscimed.2020.113534.

37. Keenan, J.M., *COVID, resilience, and the built environment.* Environment Systems and Decisions, 2020. **40**(2): pp. 216–221. DOI: 10.1007/s10669-020-09773-0.

38. Rogers, D., and E. Power, *Housing policy and the COVID-19 pandemic: The importance of housing research during this health emergency.* International Journal of Housing Policy, 2020. **20**(2): pp. 177–183. DOI: 10.1080/19491247.2020.1756599.

39. Chisholm, E., P. Howden-Chapman, and G. Fougere, *Tenants' responses to substandard housing: Hidden and invisible power and the failure of rental housing regulation.* Housing, Theory and Society, 2020. **37**(2): pp. 139–161. DOI: 10.1080/14036096.2018.1538019.

40. Garboden, P., and E. Rosen, *Serial filing: How landlords use the threat of eviction.* 2019. **18**(2): pp. 638–661. DOI: 10.1111/cico.12387.

41. Bierre, S., P. Howden-Chapman, and L. Signal, *Ma and Pa' landlords and the 'risky' tenant: Discourses in the New Zealand private rental sector.* Housing Studies, 2010. **25**(1): pp. 21–38. DOI: 10.1080/02673030903362027.

42. Melvin, J., *The split incentives energy efficiency problem: Evidence of underinvestment by landlords.* Energy Policy, 2018. **115**: pp. 342–352. DOI: 10.1016/j.enpol.2017.11.069.

43. Otter, J., *Exploitative landlord-tenant relationships in Auckland.* 2018, Auckland: Auckland Council Research and Evaluation Unit. https://knowledgeauckland.org.nz/publications/exploitative-landlord-tenant-relationships-in-auckland/

44. Gilmore, B., *'Everybody loves the landlord': Evictions & the coming prevention revolution.* Mitchell Hamline Law Journal of Public Policy and Practice, 2020. **41**(3): pp. 201–230.

45. Lister, D., *Unlawful or just awful? Young people's experiences of living in the private rented sector in England.* Young, 2006. **14**(2): pp. 141–155.

46. Beer, A., E. Baker, G. Wood, and P. Raftery, *Housing policy, housing assistance and the wellbeing dividend: Developing an evidence base for post-GFC economies.* Housing Studies, 2011. **26**(7–8): pp. 1171–1192. DOI: 10.1080/02673037.2011.616993.

47. Bentley, R., E. Baker, and Z. Aitken, *The 'double precarity' of employment insecurity and unaffordable housing and its impact on mental health.* Social Science & Medicine, 2019. **225**: pp. 9–16. DOI: 10.1016/j.socscimed.2019.02.008.

48. Baker, E., L. Lester, A. Beer, and R. Bentley, *An Australian geography of unhealthy housing.* Geographical Research, 2019. **57**(1): pp. 40–51. DOI: 10.1111/1745-5871.12326.

49. McKee, K., A.M. Soaita, and J. Hoolachan, *'Generation rent' and the emotions of private renting: Self-worth, status and insecurity amongst low-income renters.* Housing Studies, 2019. pp. 1–20. DOI: 10.1080/02673037.2019.1676400.

50. Baker, E., A. Beer, L. Lester, D. Pevalin, C. Whitehead, and R. Bentley, *Is housing a health insult?* International Journal of Environmental Research and Public Health, 2017. **14**(6). DOI: 10.3390/ijerph14060567.

51. Berry, S., D. Whaley, K. Davidson, and W. Saman, *Do the numbers stack up? Lessons from a zero carbon housing estate.* Renewable Energy, 2014. **67**: pp. 80–89. DOI: 10.1016/j.renene.2013.11.031.

52. CABE, *The value of good design. How buildings and spaces create economic and social value.* 2002, London: Commission for Architecture and the Built Environment. https://www.designcouncil.org.uk/sites/default/files/asset/document/the-value-of-good-design.pdf

53. Daniel, L., T. Moore, E. Baker, A. Beer, N. Willand, R. Horne, and C. Hamilton, *Warm, cool and energy-affordable housing solutions for low-income renters, AHURI final report no. 338.* 2020, Melbourne: Australian Housing and Urban Research Institute Limited. https://www.ahuri.edu.au/research/final-reports/338. DOI: 10.18408/ahuri-3122801.

54. Daniel, L., E. Baker, and T. Williamson, *Cold housing in mild-climate countries: A study of indoor environmental quality and comfort preferences in homes, Adelaide, Australia.* Building and Environment, 2019. **151**: pp. 207–218. DOI: 10.1016/j.buildenv.2019.01.037.

55. Simko, T., and T. Moore, *Optimal window designs for Australian houses*. Energy and Buildings, 2021. **250**: p. 111300. DOI: 10.1016/j.enbuild.2021.111300.

56. Moore, T., N. Willand, S. Holdsworth, S. Berry, D. Whaley, G. Sheriff, A. Ambrose, and L. Dixon, *Evaluating the cape: Pre and post occupancy evaluation update January 2020*. 2020, Melbourne: RMIT University and Renew. https://renew.org.au/wp-content/uploads/2020/01/Evaluating-The-Cape-research-RMIT_Renew-January-2020.pdf

57. Moore, T., F. de Haan, R. Horne, and B. Gleeson, *Urban sustainability transitions. Australian Cases – International perspectives*. Theory and Practice of Urban Sustainability Transitions. 2018, Singapore: Springer.

58. Moore, T., L. Nicholls, Y. Strengers, C. Maller, and R. Horne, *Benefits and challenges of energy efficient social housing*. Energy Procedia, 2017. **121**: pp. 300–307. DOI: 10.1016/j.egypro.2017.08.031.

59. Pevalin, D.J., A. Reeves, E. Baker, and R. Bentley, *The impact of persistent poor housing conditions on mental health: A longitudinal population-based study*. Preventive Medicine, 2017. **105**: pp. 304–310. DOI: 10.1016/j.ypmed.2017.09.020.

60. Pevalin, D.J., M.P. Taylor, and J. Todd, *The dynamics of unhealthy housing in the UK: A panel data analysis*. Housing Studies, 2008. **23**(5): pp. 679–695. DOI: 10.1080/02673030802253848.

61. Badland, H., C. Whitzman, M. Lowe, M. Davern, L. Aye, I. Butterworth, D. Hes, and B. Giles-Corti, *Urban liveability: Emerging lessons from Australia for exploring the potential for indicators to measure the social determinants of health*. Social Science and Medicine, 2014. **111**: pp. 64–74.

62. Bate, B., *Making a home in the private rental sector*. International Journal of Housing Policy, 2020: pp. 1–23. DOI: 10.1080/19491247.2020.1851633.

63. Easthope, H., *Making a rental property home*. Housing Studies, 2014. **29**(5): pp. 579–596. DOI: 10.1080/02673037.2013.873115.

64. Telfar-Barnard, L., J. Bennett, P. Howden-Chapman, D.E. Jacobs, D. Ormandy, M. Cutler-Welsh, N. Preval, M.G. Baker, and M. Keall, *Measuring the effect of housing quality interventions: The case of the New Zealand 'Rental warrant of fitness'*. International Journal of Environmental Research and Public Health, 2017. **14**(11). DOI: 10.3390/ijerph14111352.

65. Rowley, S.J., *The private rental sector in Australia. Public perceptions of quality and affordability*. 2018, Perth: Bankwest Curtain Economics Centre.

66. Cromarty, H., *Housing conditions in the private rented sector (England)*. 2021, London: House of Commons.

67. Poll, H.R., *Getting the house in order. How to improve standards in the private rental sector*. 2019, London: Citizens Advice.

68. Department for Business, Energy & Industrial Strategy, *Domestic private rented property: minimum energy efficiency standard – Landlord guidance*. 2020, UK Government, 10/03/2021 https://www.gov.uk/guidance/domestic-private-rented-property-minimum-energy-efficiency-standard-landlord-guidance

69. Ministry of Housing and Urban Development, *Healthy homes standards*. 2020, NZ Government, 10/03/2021 https://www.hud.govt.nz/residential-housing/healthy-rental-homes/healthy-homes-standards/

70. UK Government, *Private renting*. 2021, UK Government, 10/03/2021 https://www.gov.uk/private-renting

71. Burridge, R., and D. Ormandy, *Health and safety at home: Private and public responsibilities for unsatisfactory housing conditions*. Journal of Law and Society, 2007. **34**(4): pp. 544–566.

72. Dillahunt, T., J. Mankoff, and E. Paulos, *Understanding conflict between landlords and tenants: Implications for energy sensing and feedback.* In: *Proceedings of the 12th ACM International Conference on Ubiquitous Computing.* 2010, Copenhagen: Association for Computing Machinery. pp. 149–158. DOI: 10.1145/1864349.1864376.

73. Super, D., *The rise and fall of the implied warranty of habitability.* California Law Review, 2011. **99**(2): pp. 389–462.

74. Bate, B., *Rental security and the property manager in a tenant's search for a private rental property.* Housing Studies, 2020. **35**(4): pp. 589–611. DOI: 10.1080/02673037.2019.1621271.

75. Chisholm, E., P. Howden-Chapman, and G. Fougere, *Renting in New Zealand: Perspectives from tenant advocates.* Kōtuitui: New Zealand Journal of Social Sciences Online, 2017. **12**(1): pp. 95–110. DOI: 10.1080/1177083X.2016.1272471.

76. Victoria, C.A., *Repairs in rental properties.* 2021, State Government of Victoria, 10/03/2021 https://www.consumer.vic.gov.au/housing/renting/repairs-alterations-safety-and-pets/repairs/repairs-in-rental-properties

77. VCAT, *After you apply – Residential tenancy cases.* 2021, VCAT, 10/03/2021 https://www.vcat.vic.gov.au/case-types/residential-tenancies/after-you-apply-residential-tenancy-disputes

78. Raynor, K., I. Wiesel, and B. Bentley, *Why staying home during a pandemic can increase risk for some, Discussion Paper.* 2020, The University of Melbourne: Affordable Housing Hallmark Research Initiative, 06/07/2020 https://msd.unimelb.edu.au/atrium/why-staying-home-during-a-pandemic-can-increase-risk-for-some#:~:text=Confronted%20with%20high%20levels%20of,with%20higher%20risks%20of%20contagion.&text=Homeowners%20face%20other%20restrictions%20in%20their%20ability%20to%20stay%20home

79. Ahmad, K., S. Erqou, N. Shah, U. Nazir, A. Morrison, G. Choudhary, and W.C. Wu, *Association of poor housing conditions with COVID-19 incidence and mortality across US counties.* medRxiv, 2020. DOI: 10.1101/2020.05.28.20116087.

8 Government support during times of crisis

Key takeaways

- The response of governments is critical in addressing dangerous defects. Their action (or inaction) can improve (or impede) the safety, health, well-being, and social value outcomes for consumers.
- Governments typically respond to a defect crisis by tightening building regulations (attempting to avoid recurrence). They sometimes partially support the housing consumer financially but very rarely contribute 100% of the rectification costs or contribute to wider costs incurred.
- The building regulations that have been developed in recent decades have largely been centred around trying to make the construction industry 'work', with limited consideration of robust housing consumer protection.
- Building regulations need to be more consumer-focused by placing the consumer at the heart of policy thinking. This would complement current industry-based policy approaches and provide greater support and social value for consumers that encounter dangerous residential defects. This would not only help improve outcomes when a significant defect issue emerges but reduce the likelihood of them occurring, to begin with.
- Governments can create social value by providing consumer-focused policy that more comprehensively supports consumers when a human-made crisis manifests.

Chapter summary

This chapter considers how governments have responded to human-made disasters in the built environment, specifically around dangerous defects in the residential sector. Through the crisis examples of combustible cladding and leaky buildings, the way governments have responded is discussed. These examples demonstrate how the consumer is often left with significant financial costs to pay for rectification. We draw upon our research to highlight potential ways to improve consumer protection when a building crisis emerges. The need to improve consumer protection frameworks is emphasised through a case study on the 2021 Milan (Italy) fire, which demonstrates how consumers that survive potential devastating events will also need further support. The chapter concludes with recommendations for

DOI: 10.1201/9781003176336-8

the next steps, which is a complementary consumer-focused approach to building regulations. There is a six-point framework proposed for the consideration of how to move forward and help support consumers through these human-made dangerous defects that arise in the residential built environment.

8.1 An introduction to government response in the built environment

Governments around the world play a critical role in intervening and supporting their people through various disasters and crises. Earthquakes, floods, cyclones, bushfires, and the global COVID-19 pandemic are examples where government support has been required to help those affected. While there is an almost universal expectation that governments will provide support when such (largely) natural disasters occur, it becomes less clear what the role of government is when the cause of a crisis is human-made, or even potentially as a result of government action or inaction. The built environment, as opposed to the natural environment, are the human-made surroundings in which we work, live, and play. Dangerous defects within the built environment, such as asbestos, widespread leaking buildings, and combustible cladding, have caused crisis and disaster. There have been calls by those who have been impacted and wider advocates for governments to respond more comprehensively and use the range of policy and support mechanisms available to them. Within this chapter we talk about government support which can include direct financial support, financial penalties, changes to regulation and the provision of information, advice, and industry experts.

A disaster is: 'an unexpected event, such as a very bad accident, a flood, or a fire, that kills a lot of people or causes a lot of damage' [1]. Disasters that have an immediate loss of life are likely to receive intense public scrutiny and government action. For example, the fatal case examples in Chapter 1 on the 2013 Thane Building Collapse (India) resulted in some financial support for affected families and multiple prosecutions, with government legislation being breached and those responsible being found accountable. The 2021 Miami Building collapse (United States of America – see Chapter 6) could also see similar consequences, with investigations currently ongoing but with early suggestions that key internal and external governance processes were not followed or were inadequate. However, disasters do not always lead to an immediate loss of life or significant damage. Instead, outcomes can emerge more slowly and across longer time spans. This can mean it is less obvious who is accountable and in this context government action or support is often less forthcoming.

The global asbestos disaster is an example of this where 255,000 fatalities have been estimated across the world over a number of decades [2], with governments taking many years to ban the product and providing limited or no support to those affected. The asbestos issue still poses a risk, with a significant number of existing buildings around the world still containing this material. While leaving the material in place is relatively safe for the most part, it still contains a risk to health and safety if the material is disturbed (e.g. through renovation of the dwelling). Therefore, the impacts of asbestos and the need to manage its use and removal will continue long into the future. There have also been a growing number of illegal material dumping in countries, like Australia, due to the increased costs of disposal and treatment of dangerous materials (such as asbestos). This can create further and wider public health issues.

A crisis can be considered: 'a time of great danger, difficulty or doubt when problems must be solved or important decisions must be made' [3]. A human-made crisis in the built environment, such as the combustible cladding crisis, can have both immediate safety concerns and long-term health and well-being issues (see Chapter 4). Governments around the world have responded in different ways to the emergence of the flammable cladding crisis. In Australia, governments banned the use of Aluminium Composite Panels (ACPs) with a core that comprised more than 30% polyethylene in 2018. Three years later in 2021, they went further by banning materials over 7% combustible content by mass. This was not limited to polyethylene (the dangerous core in ACPs) in order to capture other materials that potentially posed a danger. The State of Victoria invested AU$600 million to help rectify the riskiest buildings, with half being directly funded by the government and the other AU$300 million being raised from a new building permit levy [4]. This funding has supported a government body, Cladding Safety Victoria, to assist affected homeowners to navigate the rectification process and provide financial assistance to those who are eligible. The state of New South Wales has taken a different approach and provided interest-free loans for affected owners, but the Australian Federal Government and other states affected, such as Queensland, have yet to provide financial support (at the time of writing). Thus, there are many homeowners across Australia that will not receive financial support, especially if they are deemed within a lower risk category building. This is of little comfort for many homeowners who have found themselves in this situation through no fault of their own.

In Europe, many countries have reviewed their fire safety regulations after the Grenfell disaster, including France, Belgium, the Netherlands, Finland, Bulgaria, Denmark, Greece Ireland, and the United Kingdom [5]. In the United Kingdom, the government responded by banning combustible materials on the external walls of new high-rise residential buildings and making it mandatory for sprinklers to be fitted in residential blocks over 11 m high [6]. A new bill, the Building Safety Bill, has also been proposed, which aims to tighten fire safety on buildings over 18 m tall. For such residential buildings 18m and over, the government will pay for unsafe cladding in England to be removed. The buildings between 11 and 18 m will have access to a low-interest loan scheme. A new developer levy that will target developers that seek permission to develop certain high-rise buildings will help fund £2 billion over the next decade. While these are useful steps, there will still be consumers of affected buildings under 18 m tall that will have to pay for rectification through a loan scheme or find another solution, such as legally challenging those responsible for the cladding. Both options are financially costly with no clear timely resolution. This can be of concern for those living in a building with unsafe cladding, which could experience a dangerous fire. The Bolton Cube Fire (see Case Study 8.1) is an example of a building that does not qualify for direct government funding but still was subject to a large fire.

Case Study 8.1   The Bolton Cube fire.

On the 15th of November 2019, a building fire at The Cube in Bolton in the North-West of England (United Kingdom) occurred around 8:30 pm [7]. The Cube building is situated near the University of Bolton and accommodates university students. Within 25 minutes, the fire had spread to all floors, putting 200 residents at risk [7]. The immediate evacuation and fire service response saved many lives, with no fatalities reported [7]. The cladding on the building was again highlighted for lack of fire safety and rapid fire-spread, leaving the building in a 'devastating condition' [7].

The fire was fuelled by a type of cladding that has not received the same scrutiny as ACPs with a polyethylene core – which has been the main focus of the cladding crisis. The cladding on The Cube was instead a high-pressured laminate (HPL) system. This is manufactured by layering sheets of wood or paper with a resin and bonding them together with pressure and heat [8]. It can be manufactured to incorporate fire retardant chemicals.

A 'standard grade' HPL is the dominant form of HPL cladding used in the United Kingdom. It is estimated to comprise 85% of all the HPL in the United Kingdom market, with 300,000 square metres sold every year to form weatherboards [9]. It has been reported that a 'standard grade' of HPL with phenolic foam insulation was tested, and failed, against a British Standards test (BS 8414) for assessing building regulation compliance [9]. Within eight minutes the test was halted after temperatures reached over 700 degrees centigrade [9]. The cladding used on Grenfell (an ACP with a polyethylene core) failed the same test in a similar timeframe [9]. Other materials, such as biowood, which is made from 70% reconstituted timber and 30% PVC (polyvinyl chloride), are also under scrutiny [10]. Thus, while the combustible cladding focus has largely been on ACPs, there are other materials that could be unsafe and be non-compliant with building compliance regulation.

It appears that homeowners in buildings that have cladding, such as HPL, which is not within the United Kingdom government funding support for combustible ACPs, will miss out on financial support. The United Kingdom government's financial support is also only for buildings that are over 18 m high. The Bolton cube is 17.84 m tall [11], meaning it would just fall short of the height criteria outlined. There will be many other buildings compromised with forms of dangerous cladding that will not receive government financial support in the United Kingdom, as they are either not ACPs with a polyethylene core or are under 18 m in height. In a report the United Kingdom Government [6] stated:

> It is clear that £1 billion will not be sufficient to remediate all 1,700 buildings with combustible non-ACM [aluminium composite material] cladding above 18 metres. The Government's own estimate is that this will cost between £3 billion and £3.5 billion. Our expectation is that the funding will only be sufficient for 600 buildings: one third of the total … We do not and should not expect the taxpayer to have to cover all remediation costs for unsafe cladding and have engaged with building owners and developers proactively to address this.

> The underestimation of rectification costs, and/or providing less financial support that will not cover all rectification work has also been seen in other jurisdictions such as Australia [12].

It is clear that the willingness of the United Kingdom government to financially assist will only go so far in the combustible cladding crisis, and there is a reliance on consumer protection regulations so that those without direct government funding have an alternative way to support themselves through the crisis. However, what is seen not just in the United Kingdom but in locations like Australia is that this wider consumer protection is failing to provide sufficient protection and is typically in favour of the construction industry rather than the housing consumer. This highlights how consumer protection frameworks must be more robust for crisis in the built environment.

Combustible cladding is not the only built environment crisis to emerge in recent years, with widespread water ingress problems also another high-profile example. The leaky home saga in New Zealand (see Chapter 2) is an example of where thousands of buildings have, and can potentially, become unhealthy through the occurrence of mould, become unsafe due to rotting timber frames, and create wider mental stress and well-being issues for homeowners as they attempt to negotiate financial and legal concerns of rectification. When such a crisis occurs, it can also affect a country's economic stability, with the leaky buildings in New Zealand estimated to cost around NZ$50 billion, or a sixth of the country's entire economy [13]. In New Zealand, the government replaced The Building Act 1991 with The Building Act 2004, which tightened regulation on the construction industry with a licensing scheme for designers, builders, and trades. Successive governments have tried to address the leaky buildings saga for owners. Attempts included the Weathertight Homes Resolution Services Act between 2002 and 2006, which aimed to provide 'provide speedy, flexible, cost-effective resolution for leaky home claims' [14] and a Financial Assistance Package scheme between 2011 and 2016, where eligible homeowners could receive a 25% contribution from government and another 25% contribution from the local council for repair work [15]. This still left eligible homeowners to pay 50% of the costs. The leaky homes crisis remains ongoing in New Zealand, despite most of the defective buildings being constructed in the 1980s and 1990s.

There are also other countries that have experienced widespread leaking buildings, such as in Canada (see Case Study 8.2).

Case Study 8.2   The leaky condo crisis in Canada.

British Colombia (B.C.) is the Westernmost province in Canada and home to over 5 million people. The coastal location of the province meant it was more susceptible to weather. In many B.C. buildings primarily built in the 1980s and 1990s, there was a significant water ingress defect that emerged, relating to rainwater infiltration into the building envelope of condominiums (condos). It has been reported that 45% of condos and 57% of schools built between 1985 and 2000 in B.C. were found to having water ingress defect issues [16].

The cladding systems that allowed water ingress in B.C. included the exterior insulation and finish system (EIFS). Theoretically, the EIFS systems should work as a barrier system with no water leakage, but this has not been the case practically [17]. When moisture has penetrated, there was then no means of escape [17], leaving the moisture in the wall cavity to cause mould and rot of timber frames, which creates structural issues over time [18]. It is now clear that EIFS requires to go beyond perfect barrier approaches, by including the provision for drainage to manage rainwater [19]. To address the leaky problems, roof overhangs were used as a solution in some cases. However, this could create other issues, as the buildings were not designed to support the additional weight of the overhangs, snow loads, and maintenance loads for inspection and repairs [20].

The affected B.C. condos were developed during a construction boom, where designers from other parts of the world, such as California, were used to keep up with demand in Canada [18]. These designers were more accustomed to designing buildings with much lower rainfall [18]. The boom also meant there was a lack of resources, resulting in many unfamiliar, inexperienced and unskilled workers contributing to the poor roofing quality in the condo crisis [21]. When the emergence of leaking buildings manifested, it was reported that there were unethical 'shell companies' being used to protect those responsible. This is where a company is found to be non-existent or insolvent when purchasers of defective housing seek recourse [22].

The former B.C. Premier Dave Barrett published a commission of enquiry report into the quality of condos in 1998. There were 82 recommendations including changes to design codes, contractor licensing, and requirements for design professionals [21]. In 1999, a new home warranty was introduced known as the '2-5-10' warranty: two years warranty on construction defects in materials and labour, five years on building envelope failures, and ten years on structural defects. The government also announced a provincial grant and tax relief program and interest-free loans to owners of leaky condos in 1999 [23]. A levy on new residential projects was introduced to support the loans. However, when a construction slowdown occurred in 2009, the B.C government discontinued the interest-free loan support, leaving owners shocked and uncertain about the future [24].

Decades on, the condo crisis is an ongoing construction, legal and financial crisis in Canada. In 2001, the B.C. Government estimated the cost to repair leaky condos was CA$1.5 billion [23]. This cost will only have increased since, but there has since (to date) been no financial assistance from the federal government.

In Canada, the government took some form of action, including tightening of construction and buildings regulations to avoid an occurrence of similar issues emerging in the future. While this is a good step, the reaction is largely centred around tweaking or reforming the construction sector, and thus placing the industry at the centre of policy and legislative changes and considerations. It also does little to address the issue that was created in the first instance. This leaves the consumer still lacking robust protection.

Another example includes the home insulation program from Australia during 2009–2010, which is more commonly referred to as the 'Pink Batts' disaster due to the colour of the ceiling insulation used. It was widely seen as a disaster by households and key industry/policy stakeholders in Australia due to a range of defect and safety issues which emerged in the short-lived program for installers and housing consumers. This included the deaths of four installers and more than 200 house fires (mainly from faulty wiring and installing the insulation too close to down lights) before the program was abruptly discontinued [25]. Across the wider housing community, these house fires created significant concern that those who had been part of the program had increased the risk of their house catching fire.

This is an example where issues by both the government and the construction industry contributed to a significant defect outcome [26]. The program was designed to provide a rebate of up to AU$1600 to 2.2 million owner-occupied existing homes to insulate their roof space and by doing so reduce heating and cooling energy use by up to 40%, reduce associated greenhouse gas emissions and generate employment for the building industry [27, 28]. Before the program was started, an independent risk assessment of a trial of the program found that it should be delayed by at least three months due to several risks with governance and implementation of the program, which would potentially place workers and households at risk of negative outcomes. However, the program was rolled out with only a few additional tweaks. Almost immediately the program was plagued by issues around changing standards, issues with rebate payments, rogue installers, installers with limited experience, and product safety concerns [27, 28].

Two of the deaths were directly related to the use of foil insulation. A later review found that the inclusion of this insulation material was 'fundamentally flawed' and that risk advice had been received before the program began [25]. While the program provided insulation to 1.2 million dwellings and achieved many of the wider benefits aimed for, the issues with the program resulted in more than 200,000 dwellings needing to have their insulation checked to ensure safety for households. This was an additional cost for the government of more than AU$425 million. There is also a lasting legacy across the construction industry, housing consumers, and government in relation to insulation and large-scale programs which try to drive improvements of quality and performance.

The above insulation example, and those of combustible cladding examples in the United Kingdom and Australia, and the examples of the leaky building's crises in New Zealand and Canada has demonstrated that housing consumers have not been comprehensively considered by governments and other key stakeholders, leaving wider implications on safety, health, well-being, and social value. While some of the worst-affected, or riskiest, buildings may have received some financial or rectification support from governments, many homeowners are left without adequate assistance [12, 29]. It is argued in this chapter that a more effective approach would be to place the consumer at the centre of policy and legislative changes. Thus, rather than only asking reactive questions centred around the industry, such as how we can fix the industry to stop another crisis, other proactive questions must be considered, including: how can we support consumers when crisis occurs? If we place the consumer at the centre of policy thinking, we begin to ask these more pertinent questions that complement the industry-focused policy approach. This would create a broader perspective (as well as a much more proactive approach) that assumes crisis will occur, rather than reactively trying to fix an industry, when it has not worked as intended. This would ensure there is a greater focus on the housing consumers, and that they are not left 'high and dry' when issues do emerge. After all, for many housing consumers their dwelling will be the most expensive thing they ever purchase; yet it is in many jurisdictions the thing they have least consumer protection for. This imbalance must be addressed.

The following section presents our research on the perspective of placing consumers at the centre of policy and regulation, through exploring ways to help support consumers affected by crisis in the built environment, before discussing next steps into considerations for placing the consumer at the heart of regulation.

8.2 Considerations for how to help the consumer in crisis

The consumer is often left without comprehensive protection and support when crisis emerges within the built environment. The examples of the leaky building crisis and the combustible cladding crisis discussed above show how governments typically respond by tightening building regulations to avoid recurrence (or new similar issues) and offering some financial or rectification process assistance to homeowners. This assistance has typically not covered the full financial rectification costs of the defect for all homeowners, with many receiving no financial help or a percentage contribution. As highlighted in the previous chapters, financial costs also go way beyond the cost of replacing the physical defect, with building insurance and other costs emerging from crisis. There is also the mental health cost and implications for occupant well-being and social value. Typically, these wider costs are not covered within any rectification financial support provided by governments.

Our research explored how the consumer could be more comprehensively supported through their perspectives and lived experience during the crisis [12, 29]. This research on the cladding crisis, can be broadened to other dangerous defects that: pose health and safety concerns, threaten the well-being of residents, and erode the social value they place on their homes. In our cladding research, homeowners wanted to know if they were safe or what could be done immediately to ensure they were safe. They wanted action quickly, even if it was a temporary measure that reduced risks, such as increased use of sprinklers. For instance, one homeowner stated:

> I would like in my building sprinklers on the balconies, immediate, effective tomorrow orders. We don't care how you do it. You're going to put sprinklers on those balconies because it could take years for this cladding issue to be sorted out. That would be my deepest wish for my personal property, is sprinklers on the balconies … If the building's in flames, it's going to cost a good deal more than what sprinklers on the balcony are going to cost.

As well as interim fire safety measures, homeowners also wanted some transparency about a timeline for when buildings were going to be assessed and rectified:

> I would like to see some realistic timelines with respect to how long Cladding Safety Victoria believes they'll be able to go through different buildings … It

could take five, 10 years or so. But if you have a guarantee there that says, in the process, if you can put up some interim measures, then we will be satisfied until the building is properly assessed and properly remediated.

They also thought that governments bodies that are dealing with cladding should be better resourced to help answer support the homeowners, especially given the urgency many homeowners felt about getting clear information to at least put their own minds at ease:

At the moment, well, what I would like to see is basically a lot more resourcing staff resourcing in government agencies to deal with it, because it seems like they're completely overwhelmed. And so therefore, we're having to wait for very long periods of time to get any assessment and any advice.

Where possible homeowners also requested what rectification designs (e.g. full removal of cladding or part rectification) would suffice to have their government building notices removed. For example, one homeowner stated:

In an ideal world, I would like to see Cladding Safety Victoria come out with a transparent discussion about what it is that they deem to be the qualities in the building that they deem to be safe to have the [building] notices removed … So, you actually know how much money you will ultimately be liable for to resolve the issue … So a bit more transparency about what they would consider to be safe.

With homeowners facing potentially high financial costs, there were also requests for transparency around the government financial support in the State of Victoria. This was regarding who would receive financial support and how much they would get. While homeowners were aware the riskiest buildings would receive funding, many buildings had unclear or unknown risk ratings, meaning it was uncertain who would financially benefit. One homeowner stated they wanted:

A bit more information about the money that the state government has and how it will be spent and to whom, and which buildings will receive it and how much of it they'll receive.

Though in terms of moving forward, homeowners believed they should all receive financial help and ideally cover as much of the rectification work as possible. For example, one homeowner stated:

I'd like to see us be eligible for government money, for the remediation. I was quite excited to hear about an alternative method of treatment that could be about wrapping the buildings rather than taking cladding off and replacing it, which is hugely expensive. So, I mean they're very … It's a dream list really. All those things and whether it will be solved in our case, either with government funding or [the builder] will say 'we're very sorry, here's a load of money'.

Another homeowner believed that it was only fair that their building (which had AU$1 million worth of rectification work) would be fixed at the builder's cost and not the taxpayers:

I'd like to see us be given the million dollars to come in and fix the building by the builder, not by the taxpayer. And I know that the builder has the assets, whether in his name or not is another thing. But that builder is still building. I'd like to see our building fixed by the builders giving us the cash to fix it and him not fixing it. I like to see him de-registered and I'd like to see all his assets actually frozen, all the land that he still owns, and he cannot build on those properties.

It was also suggested that a mix of approaches was appropriate, with taxpayers initially fronting the bill, only for them to try and recover the money from those in industries that were accountable:

I think the Cladding Safety Victoria should pay for it, and it's up to them to recover the money. Like they said in a press release, that, or the State Government said 'We'll help, and re clad all these buildings, and we'll go after the perpetrators'. That's what I want to see, and I don't think that's the ideal world, I think that is what exactly is the correct thing to do … The government … they're the only ones that have got the resources to stump up and fix this, and then they will recover it from the people that have done the wrong thing in creating it in the first place, but the owners should be indemnified. There shouldn't be any cost effect on the owners because it's absolutely not their fault.

It should be noted that while these requests were directed towards the Victorian government, other state governments in Australia were yet to move on the matter:

And we're hoping to work with the Australian Capital Territory government to identify what's going to happen. I mean, Victoria's the model, where the government has said that they're prepared to assist unit complexes, but our government is silent at the moment.

These example quotations from our research results highlight some of the considerations for moving towards a more consumer-focused built environment. These included providing: immediate fire safety measures to reduce risks, government resources to advise consumers, a clear outline of the process to rectification, and financial support for construction works. For comprehensive consumer protection, the support frameworks in the built environment should not only be restricted to these actions. There is also further consideration required for capturing all types of dangerous defects, as well as support for homeowners that are affected by dangerous defects more than others, such as survivors of a combustible cladding fire (see Case Study 8.3).

Case study 8.3   Milan fire: A case for further supporting survivors.

On the 29th of August 2021 a fire in Milan (Italy) occurred on a 20-storey apartment tower block, called Torre dei Moro in Milan [30]. In just over three minutes, the entire façade of the building was in flames [31]. It was reported that the flammable material on the façade was produced by a company that publicly (on their website) do not recommend the use of the material for skyscrapers [32]. The rapid spread of the building fire was yet another example of the risks of combustible materials on high-rise facades.

The building was home to 70 families, with approximately 30 people in the building when the fire emerged [33]. There were 20 that suffered smoke inhalation, but all were safely evacuated during the fire rescue mission. The survivors were temporarily housed in hotel accommodation. It was reported that the city did offer to pay for the accommodation for one month, though hotels were paid for by a building sinking fund for maintenance [34]. However, survivors had to support themselves with their own resources beyond that. Their future at the time of writing is uncertain, with survivors explaining that the new accommodation that has been proposed for them is far from their previous home, schools, work, and social networks [35]. Survivors of the Grenfell Tower disaster have also reported similar challenges of being stuck in unsuitable temporary accommodation, even more than three years on [36].

The Torre dei Moro survivors have spoken about how they feel alone and empty after their loss of all everything they owned in the fire including their wallets, phones, and furniture. One resident stated [33]:

> We had invested everything for that apartment… Everything we had went up in smoke. We have nothing left.

Survivors were also concerned that the building, Torre dei Moro, would risk stigma and loss of property worth even after repairs [35]. They have been advised that they will have the opportunity to buy new properties with a small discount in February 2022 (five months after the fire) [34]. However, these properties are above the market average (even with a discount), meaning many residents could not afford them [34]. Survivors believed they were receiving little notice or attention, despite their fate being in the hands of the government [34].

Overall, there is a clear need for greater consideration for supporting those affected by dangerous defects. In current frameworks, governments have acknowledged themselves that support will take time, and funding is limited or not forthcoming. For example, Cladding Safety Victoria stated:

> There are practical constraints on the number of buildings that can be rectified at any one time, which means that the project overall is expected to take five years to complete. These constraints include limited government funding, the availability of fire engineers and other necessary professionals, the requirement to go through approvals processes, including the Building

Appeals Board, and the limited resources of Cladding Safety Victoria in assessing hundreds of individual projects [37].

There are other governments in Australia that are far less active than Cladding Safety Victoria, and the government in the United Kingdom have taken many years to formulate their response to the ongoing cladding crisis. The current outcomes for consumers with dangerous defects are far from ideal. There is a clear need to consider re-thinking and re-framing policy to deliver a more consumer-focused approach than current approaches, which is further discussed below.

8.3 Next steps: Complementing with a consumer-focus

The current building regulations in many countries across the world largely revolve around the industry stakeholders in construction and ensuring a minimum quality for the housing consumer. In Australia, the National Construction Code and various Australian Standards focus on making the construction industry 'work' through policies and regulation [38, 39]. Similarly, in Europe the Eurocodes are used, and in America, the International Building Codes and International Residential Code were established by the International Code Council [40]. These various codes and standards cover a wide range of construction topics, such as the building of structures, mechanical services, plumbing, and energy conservation. It is of course essential that the construction industry is regulated in an attempt for buildings to function appropriately and to ensure there is a minimum quality and performance for housing consumers (that regulators, industry, and housing consumers deem acceptable). This in turn can lead to better outcomes for consumers, though various construction crisis events, such as asbestos and combustible cladding, have highlighted that more work is required in this space, and despite best intentions, issues still occur. In particular, it has been reported that the deregulation of the building industry in recent years has contributed to major defects [41] by allowing greater flexibility without sufficient accountability [42]. There is clearly still significant policy work to be undertaken to ensure high-quality buildings are made that create social value and enhance occupant well-being.

Beyond a focus on making the construction industry 'work' through more robust building regulations, there is also a need to make consumer protection 'work' within the built environment. A greater consumer-focus on policy and regulation would complement the current industry-based frameworks. Current frameworks for consumer protection in the built environment typically cover new residential property for defects for up to (approximately) ten years, depending on jurisdiction. See for example the '2-5-10' warranty in B.C. (Canada) in the case example above. However, when defects emerge there are still many current challenges for accessing this building warranty protection, such as:

- construction companies not responding to defect complaints;
- legal action is very costly and can take many years;
- multiple construction stakeholders, meaning it is difficult to pin-point who is responsible to fix the defects;

- construction companies may have dissolved and/or 'phoenixed';
- international construction supply chains, meaning it can be more difficult to provide warranty for defective products from overseas; and
- warranties have time-restrictions and clauses (e.g. cladding requires cleaning every six months) for approval.

It is argued throughout this book that there needs to be more of a consumer-focus within the built environment and, in particular, for dangerous defects. These types of defects are: 'A major shortfall in the building performance that emerges after the building is in use, which is not rectified in a timely manner, is costly to address, and poses a continuous risk to the occupants' safety, health or well-being, that can last for years' (see Chapter 2).

Asbestos, flammable cladding, and widespread leaky buildings are examples of dangerous defects. Since they fall on the major side of the severity scale it is argued that a carefully considered disaster relief policy framework should be formulated for such dangerous defects. Considerations and ideas for a more comprehensive consumer-focused framework based on the research presented in this book are presented in Figure 8.1.

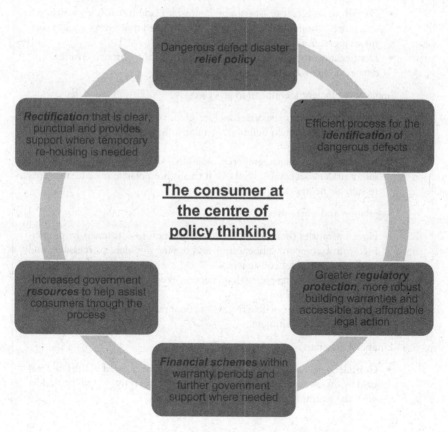

Figure 8.1 Example of policy-thinking, with the consumer at the centre.

Figure 8.1 is an example of policy-thinking that places the consumer at the heart. Current building policy thinking largely centres around the industry. By considering the consumer at the heart of policy there could be more robust protection for dangerous defects (and addressing minor defects as well). The first stage would be for greater clarity on when disaster relief policy can extend beyond natural hazards (e.g. floods) to human-made hazards in the built environment (e.g. use of asbestos). The definition for what qualifies as a human-made disaster would need to be categorised clearly, hence the dangerous defect definition we proposed in Chapter 2. Dangerous defects that emerge will require efficient identification and action, as they are an immediate health and safety concern. To appropriately respond, robust consumer protection is required: with more comprehensive and nuanced finance schemes, increased resources for support, advice, and timely execution of the building rectification. The six points presented in the diagram above for consideration are further elaborated on in Text Box 8.1.

Text Box 8.1   Elaboration of policy-thinking that places the consumer at the heart.

1 Initiating dangerous defect relief support

- Dangerous defects (e.g. asbestos or flammable cladding) are clearly defined, so a process for dangerous defects can be initiated that brings greater consumer protection.
- Identification should be from an independent third party to ensure confidence and reduce conflict of interest.

2 Implement efficient identification processes

- Building material passports could be used, so it is quickly known what materials are in which buildings (similar to how ingredients are disclosed in food products).
- Implement immediate temporary measures that reduce any health and safety risks to acceptable levels, or if this is not possible relocate residents to suitable nearby accommodation.

3 Regulation and warranties

- Have warranties that provide greater protection to consumers by protecting against company 'phoenixing' and having affordable, accessible, and swift legal action for consumers.
- Consider also having policy that encourages companies to act with corporate social responsibility.
- Require stakeholders with a history of poor practices to provide longer warranties than the minimum.

4 Financial Schemes

- Consider defect schemes, where an amount of money owed to the builder is held by the owners corporation or government and only paid to the builder once the warranty period ends (if defect free).

- Have government support packages available for rectification work where required but ensure it can provide for all impacted owners.
- Consider packages of support for other financial costs, such as building insurance rises and homes becoming unsellable, as well as human costs, including mental health support.
- Ensure that any other impacts, such as increased insurance costs during the period where the defect is not yet rectified, are reduced/addressed to account for new building safety.

5 Resources

- Provide increased government resources that help advise and support all types of consumers, owners corporations, strata and building managers and increase resources during times of crisis to ensure timely provision of support and information.
- Utilise industry experts where needed for solutions.
- Outline a clear and transparent process with milestones and timelines.

6 Rectification

- Clearly communicate designs and work required to make buildings safe and process to be removed from any 'dangerous buildings' lists.
- Provide additional support for those requiring to be re-housed during rectification.
- Undertake works in a timely fashion for consumer well-being.

8.4 Conclusions

In the event of a large-scale human-made crisis in the built environment, such as asbestos, widespread leaky buildings, and combustible cladding, consumers require support from the government. The rectification of dangerous defects is currently a long process, due to the scale of such crisis and lack of robust consumer protection. The current building regulations are industry-focused and attempt to regulate the construction industry to create buildings of the required standard. There are clearly flaws in the current regulatory systems, as crises in residential buildings have manifested. These building regulations need to be made more robust. However, there also needs to be a much greater focus on the consumer to complement the current building regulations.

By placing the consumers at the centre of policy-thinking, the response to emerging crisis in the built environment would be much more comprehensive. The central question that should be asked is: how can we support consumers when dangerous defects arise? The answer to this question involves: initiating disaster relief support, identifying all buildings affected quickly and efficiently, providing adequate consumer protection regulation, using policy to encourage building companies to act with corporate social responsibility, making it possible for consumers to access and afford legal advice and action, implementing financial schemes that are accessible in the event of defects, providing adequate resources to deal support homeowners through a clear and transparent rectification process, and provide additional support to those who have been most affected by the crisis, such as those needing rehoused.

The sooner that the consumer is placed at the centre of policy-thinking, the sooner consumers will be better supported through a built environment crisis. A dangerous defect that is rectified quickly, with a clear process, at no cost to the consumer when under warranty, will improve health, safety, well-being, and social value outcomes.

References

1. Oxford Learner's Dictionary, *Disaster*. 2021, Oxford University Press, 10/13/2021 https://www.oxfordlearnersdictionaries.com/definition/american_english/disaster
2. Furuya, S., O. Chimed-Ochir, K. Takahashi, A. David, and J. Takala, *Global asbestos disaster*. International Journal of Environmental Research and Public Health, 2018. **15**(5): p. 1000.
3. Oxford Learner's Dictionary, *Crisis*. 2021, Oxford University Press, 10/8/2021 https://www.oxfordlearnersdictionaries.com/definition/english/crisis_1
4. Andrews, D., *Tackling high-risk cladding to keep Victorians safe*. 2019, Victorian Government, 10/13/2021 https://www.premier.vic.gov.au/tackling-high-risk-cladding-keep-victorians-safe
5. Fire Safe Europe, *A year of changes: Taking stock of the evolution of fire safety requirements in buildings throughout Europe*. 2018, Fire Safe Europe, 10/14/2021 https://firesafeeurope.eu/year-of-changes-evolution-of-fire-safety-requirements-in-buildings-throughout-europe/
6. UK Government, *Government response to the housing, communities and local government select committee report on cladding: Progress of remediation*, 2020. London: Crown Copyright. https://www.gov.uk/government/publications/cladding-progress-of-remediation-government-response-to-the-select-committee-report
7. BBC News, *The Cube: Bolton student flats blaze evacuation 'saved many lives'*. 2020, British Broadcasting Corporation, 10/14/2021 https://www.bbc.com/news/uk-england-manchester-53597167
8. Anonymous, *HPL cladding*. 2021, Designing Buildings – The Construction Wiki, 10/14/2021 https://www.designingbuildings.co.uk/wiki/HPL_cladding
9. Apps, P., *Widely used HPL cladding system dramatically fails official fire test*. 2020, Inside Housing, 10/14/2021 https://www.insidehousing.co.uk/news/news/widely-used-hpl-cladding-system-dramatically-fails-official-fire-test-65870
10. Chung, F., *'Undue fire risk': Wood cladding used on hundreds of buildings including McDonald's ruled 'combustible'*. 2019, Nationwide News Pty Ltd, 10/14/2021 https://www.news.com.au/finance/business/other-industries/undue-fire-risk-wood-cladding-used-on-hundreds-of-buildings-including-mcdonalds-ruled-combustible/news-story/05a8ce3990aeaf69ce45747015409d54
11. UK Government, *Cladding: Progress of remediation*, 2020, London: House of Commons, UK Government. https://publications.parliament.uk/pa/cm5801/cmselect/cmcomloc/172/17203.htm
12. Oswald, D., T. Moore, and S. Lockrey, *Combustible costs! Financial implications of flammable cladding for homeowners*. International Journal of Housing Policy, 2021. pp. 1–21.

13. Dyer, P., *Rottenomics: The story of New Zealand's leaky buildings disaster*. 2019, David Bateman Limited.

14. NZ Government, *Introduction to weathertight homes tribunal*. 2015, NZ Government. https://www.justice.govt.nz/assets/Documents/Publications/WHT-introduction-to-wht.pdf

15. NZ Government, *Financial assistance package for weathertight claims*. 2019, NZ Government. https://www.building.govt.nz/resolving-problems/resolution-options/weathertight-services/financial-assistance-package-for-weathertight-claims/

16. Weslowski, K., and M. Thomson, *Gone but not forgotten – Water ingress claims in British Columbia: Will rainscreens and building envelope professionals prevent another "Leaky condo crisis"?* 2016, Vancouver: Miller Thomson LLP.

17. Koester, J., *Claddings and entrapped moisture: Lessons learned from early EIFS*. 2013, The Construction Specifier. https://www.constructionspecifier.com/claddings-and-entrapped-moisture-lessons-learned-from-early-eifs/

18. Baerg, M., *How to avoid buying a leaky condo in Greater Vancouver: The basics of leaky condos & rainscreening*. n.d., Bridgewell Real Estate Group. https://bridgewellgroup.ca/leaky-condo/

19. Lstiburek, J., *BSD-146: EIFS – Problems and solutions*. 2007, Building Science Corporation. https://www.buildingscience.com/documents/digests/bsd-146-eifs-problems-and-solutions

20. Lazaruk, S., *B.C. condos built to fix leaky problem with overhangs now facing bigger problem*. 2020, Vancouver Sun, 10/11/2021 https://vancouversun.com/news/local-news/condos-built-to-fix-leaky-problem-with-overhangs-now-facing-bigger-problem

21. Barrett, D., *Commission of inquiry into the quality of condominium construction in British Columbia, chapter 2: The framework of residential construction*. 1998, Nanaimo: Government of British Colombia.

22. British Columbia Law Institute, *Interim report of the project committee on new home warranties*. 2000, Vancouver: British Columbia Law Institute. http://www.bcli.org/sites/default/files/NHW_Interim.pdf

23. Penner, D., *Leaky condo crisis rears its head again in B.C.* 2014, Vancouver Sun, 10/11/2021 https://vancouversun.com/business/real%20estate/leaky-condo-crisis-rears-its-head-again-in-british-columbia#:~:text=Evidence%20of%20a%20second%20wave,depreciation%20reports%20on%20their%20buildings

24. CBC News, *B.C. leaky-condo loan fund dries up*. 2009, CBC News, 10/11/2021 https://www.cbc.ca/news/canada/british-columbia/b-c-leaky-condo-loan-fund-dries-up-1.839054

25. Hanger, I., *Report of the Royal Commission into the Home Insulation Program*. 2014, Canberra: Australian Government. https://apo.org.au/node/41087

26. Moore, T., *Strategic niche management and the challenge of successful outcomes*. In: T. Moore, et al., ed. *Urban sustainability transitions. Australian cases – International perspectives*, 2018. Singapore: Springer.

27. Hawke, A., *Review of the administration of the Home Insulation Program*. 2010, Canberra: Australian Government.

28. Australian National Audit Office, *Home Insulation Program*. 2010, Canberra: The Auditor-General. https://www.anao.gov.au/work/performance-audit/home-insulation-program

29. Oswald, D., T. Moore, and S. Lockrey, *Flammable cladding and the effects on home-owner well-being*. 2021, Housing Studies.

30. Summers, A., *Grenfell-like fire at tower block in Milan, Italy endangers hundreds of lives*. 2021, World Socialist Web Site, 10/08/2021 https://www.wsws.org/en/articles/2021/09/05/mila-s05.html

31. Verzoni, A., *Totally vulnerable*. 2021, National Fire Protection Association. https://www.nfpa.org/milan

32. Deborah, *Fire in via Antonini in Milan, first investigated*. 2021, Buongiorno Milano Newsletter, 10/13/2021 https://www.italy24news.com/local/215119.html

33. France-Presse, A., *Fire rips through 20-storey residential tower block in Milan*. 2021, The Guardian, 10/08/2021 https://www.theguardian.com/world/2021/aug/29/fire-rips-through-20-storey-residential-tower-block-in-milan

34. Apps, P., *'They are trying to minimise it as a private accident covered by insurance': The story of a giant cladding fire in Milan*. 2021, Inside Housing, 10/08/2021 https://www.insidehousing.co.uk/insight/they-are-trying-to-minimise-it-as-a-private-accident-covered-by-insurance-the-story-of-a-giant-cladding-fire-in-milan-72743

35. Norris, S., *What a blaze in Milan can tell us about the Grenfell fire*. 2021, Byline Times, 10/08/2021 https://bylinetimes.com/2021/09/21/what-a-blaze-in-milan-can-tell-us-about-the-grenfell-fire/

36. Apps, P., *Grenfell survivor still waiting for rehousing three years on speaks of 'nightmare'*. 2020, Inside Housing, 10/08/2021 https://www.insidehousing.co.uk/news/news/grenfell-survivor-still-waiting-for-rehousing-three-years-on-speaks-of-nightmare-67129

37. Cladding Safety Victoria, *About cladding safety Victoria's program*. 2020, Victorian Government, 10/14/2021 https://www.vic.gov.au/find-out-about-cladding-safety-victorias-program

38. Moore, T., and S. Holdsworth, *The built environment and energy efficiency in Australia: Current state of play and where to next*. In *Energy performance in the Australian built environment*, 2019. Springer: Singapore.

39. Berry, S., and T. Marker, *Residential energy efficiency standards in Australia: Where to next?* Energy Efficiency, 2015. **8**(5): pp. 963–974.

40. Moore, T., R. Horne, and J. Morrissey, *Zero emission housing: Policy development in Australia and comparisons with the EU, UK, USA and California*. Environmental Innovation and Societal Transitions, 2014. **11**: pp. 25–45.

41. James, B., M. Rehm, and K. Saville-Smith, *Impacts of leaky homes and leaky building stigma on older homeowners*. Pacific Rim Property Research Journal, 2017. **23**(1): pp. 15–34.

42. May, P., *Performance-based regulation and regulatory regimes: The saga of leaky buildings*. Law & Policy, 2004. **25**(4): pp. 381–401.

9 Designing and building better for the housing consumer

Key takeaways

- Previous chapters have discussed the reactive response to building defects. This chapter demonstrates how proactive thinking to building quality is required to both reduce risks of future defects and improve social value through design and construction of housing.
- There are an increasing number of examples around the world which demonstrate a range of enhanced social benefits such as improved dwelling quality, design for disassembly, reduced living costs, and improved occupant health and well-being.
- There is no technical or material reason this type of housing could not be provided for everyone, but it does require a shift in focus by housing consumers, the construction industry and governments.
- Importantly, enhancing social value in housing must include delivering such outcomes for those who are most vulnerable in our communities (e.g. low-income households) to ensure an equitable transition to improve building quality and performance.
- Focusing on enhancing social value in housing to begin with will also likely mitigate issues of minor and major defects.

Chapter summary

Much of this book is focused on how the building industry, housing consumers, and governments have responded once minor or major issues of quality, defects, performance, and crisis manifest and the implications this has on social value. This chapter pivots to look at how we can enhance social value in housing, to begin with, and in doing so provide a range of positive social value outcomes for housing consumers, the construction industry and governments. The chapter starts by exploring the evidence around how improved design and construction can lead to better consumer benefits, such as lower utility bills and wider societal benefits, such as sustainability. This is followed by several

DOI: 10.1201/9781003176336-9

case studies, including from our own research, where this has occurred. We follow this with a detailed discussion on how better building design and construction can lead to homes that consumers socially value, as well as reflections on how the various roles government, industry, and consumers have for driving change.

9.1 Introduction to delivering improved social value

The preceding chapters in this book have largely focused on consumer, industry stakeholders and government responses when defects, disasters, and crises have already emerged within the residential sector. This response has typically been in two parts: (1) working to rectify the issues of any impacted buildings and (2) revising policy and industry practices to mitigate the chance of the defect, disaster or crisis occurring again. Beyond these typical responses, it is important to consider raising the minimum standards of buildings to a significantly higher quality that will further reduce the likelihood of dangerous defects occurring and create greater health, well-being, and social value outcomes for consumers. There will also be a range of wider benefits of higher quality buildings, such as improved environmental sustainability of housing which will help to contribute to outcomes of another crisis, the climate crisis, which is becoming increasingly urgent.

While it is critical that policy makers, the building industry, and housing consumers continue to focus on addressing the range of issues and negative social value seen across the residential sector, we must also recognise that there is an opportunity, and even moral obligation, to deliver enhanced social value in our housing and across our cities. This is even captured within the United Nation's Sustainable Development Goals. For example, Goal 11 is all around making cities and human settlements inclusive, safe, resilient, and sustainable. Importantly, we know from the wider research that improved housing design, quality, performance, sustainability, and liveability can lead to improved social value outcomes for housing consumers, the wider community, industry, and policy makers [1–18]. This includes outcomes such as reducing the frequency and types of defects (both major and minor), extending the quality and life of the dwelling, reducing living costs and prevalence of energy poverty, improving thermal comfort and occupant health and well-being. The increasing body of evidence and case studies from around the world demonstrates that housing, which enhances social value, is not only possible, but it can also be done for little, if any, additional costs for design and construction. Furthermore, the evidence shows that enhanced social value is not limited just to the occupants of the dwelling but cascades throughout communities leading to enhanced community social value, with benefits also for industry and government. The following sections explore some of the research that has explored how we can enhance social value in our housing and why we need to focus on including this as a starting point for new housing and retrofitting of existing housing.

Before we discuss some of the evidence about enhancing social value, we should first reflect on what type of dwelling might enhance social value. While this discussion will be geographically and culturally dependent and is also continuing to change across time, there are some common themes we can highlight:

- A dwelling that can enhance social value is one which has been designed and constructed to not only meet the basic needs of the household but has included opportunities to address other needs.
- Importantly, it is a dwelling which leads to a significantly improved level of performance and quality.
- It is a dwelling which when constructed is free from defects and is able to maintain a high quality of finish and durability across the life of the dwelling and can be easily maintained as the dwelling begins to wear.
- It is also a dwelling which has reduced operational impact (e.g. reduce energy and water consumption, low maintenance requirements) and reduces the impact on the environment (e.g. low embodied energy in materials, low impact from operation, design for end of life).
- It is also able to improve health and well-being and other social outcomes for occupants.

There are an increasing number of examples of dwellings that deliver this type of outcome and the evidence about the benefits from such housing continues to grow.

There are a variety of ways in which a well-designed, constructed, and performing dwelling can enhance social value for households (or occupants). A dwelling built with a focus on overall quality, performance, and liveability and, without shortcuts, will reduce the likelihood of minor or major defects occurring (and in fact should be defect free to begin with). The enhanced social value here is the absence of any of the stress, anguish, and financial implications described in earlier chapters when homeowners go through the process of having minor or significant defects rectified. Given the earlier examples discussed in this book, such as leaky homes in New Zealand (see Chapter 2) and the structural damage to the Downtown Sarasota apartment in the United States, it can take many years, if not decades, for some of these larger-scale defect issues to be addressed. Therefore, the absence of defects is not just an enhanced benefit at the point of time of taking ownership or occupation of a dwelling but continues through the life of the building.

Furthermore, while not typically considered a defect, the issues of poor quality housing are gaining increasing attention. Poor quality housing is housing which may not have obvious defects as they are traditional considered by policy makers, the building industry, and consumers but are things which on their own, or as an accumulation, mean the dwelling and household are not able to maximise outcomes of living in the dwelling. A key example of this is a thermally poor envelope (e.g. from a lack of insulation) which leaves houses cold in winter and warm in summer and takes significant amounts of energy to heat and cool, meaning higher financial costs for consumers.

There is a significant and growing body of research around the world which demonstrates the relationship (both positive and negative) between housing quality, design, and performance and the implications on the economic and social well-being of housing consumers, as well as on the wider natural environment [7, 19–24]. This has been exacerbated during COVID-19 with the quality of peoples' dwellings impacting their ability to not only stay safe but also to undertake other activities such as working from home and mental and physical health and well-being [25–36]. The COVID-19 pandemic has reframed the way people think about housing. It is now often thought of as more than a place to sleep, but it is a place that can enhance liveability, where occupants can have high levels of social value placed on their home, and it has revealed pre-existing inequalities, exclusions, and deprivations based not only upon housing quality but also household demographics [26, 28, 32, 34–39].

There is increasing global research emerging which explores the health and well-being implications for occupants in various types and quality of housing. In particular, research has found that improved quality, thermal performance, and affordability of a dwelling can significantly improve real and potential liveability and health implications for households (and conversely that poor quality and thermal housing negatively impacts occupants' liveability, health, and well-being) [2, 5, 6, 40–52]. For example, significant health benefits such as a reduction in respiratory disease, improved sleep, reducing the severity of issues like arthritis, and a general reduction of colds, coughs, and other milder ailments have been found [2, 6, 20, 22–24, 40, 50]. These health benefits are often more significant for those who are most vulnerable in our communities such as children, the elderly, or those who are low-income. For example, one evaluation of the benefits to health from improved design and sustainability (including improved indoor air quality) of 37 public housing tenants in Boston (United States) found that in comparison to standard (design and construction) houses, the households in the improved quality and sustainable housing experienced a reduction in self-reported health issues of 57% [53]. Furthermore, the health implications and cost for individual households add up to a significant wider social cost. Research from the United Kingdom has found that the cost of people living in the bottom 15% of United Kingdom housing cost the National Health Service £1.4 billion per year [52].

Improvements to the quality and thermal performance of a dwelling do not just improve occupant health and well-being but also result in reduced requirements for mechanical heating and cooling (both in terms of needing smaller sized systems, or eliminating them altogether, as well as a reduction in how much they are used) [9, 54]. This reduces capital and operating costs which helps improve living affordability for households. There are an increasing number of examples around the world where a combination of improved dwelling quality and performance (including onsite renewable energy and rainwater collection) has resulted in eliminating day-to-day bills related to energy and water costs. The increasing move towards 'smart' homes is promising to elevate these benefits even further and place the dwelling and householder at the centre of a dynamic two-way interaction with the wider urban and energy environment [11, 55, 56].

In some regions of the world, improved design, construction, and use are saving households several thousands of dollars a year. In a climate of rapidly increasing utility costs, this is making such housing more affordable from a through-life liveability perspective. Research from Australia found that for a high-performing zero-energy dwelling the household could save more than AU\$90,000 in energy bills over the assumed 40-year life of the dwelling [57]. Further, if energy savings were re-invested back into mortgage repayments the house would more than offset any additional capital cost borrowing and result in a reduction of interest paid on the home loan of more than AU\$50,000 (paying off the house up to four years sooner). This is not just applicable for new housing but even undertaking even small retrofitting works such as draught sealing, installing ceiling fans, and internal changes to prevent heating and cooling of service spaces could make a significant reduction to energy costs for tenants [58].

Dwellings that help reduce living costs *and* improve amenity and performance are increasingly important for the growing cohort of households who are in, or near, fuel/energy poverty in many regions [59–63]. Fuel poverty, where occupants cannot afford to pay for sufficient energy to meet basic living requirements (such as maintaining thermal comfort within a health range), has been identified as a significant issue in places such as the United Kingdom, Europe, and Australia [63, 64]. Fuel poverty is not just about the economics of paying for energy consumption but also leads to increased health issues through an inability to maintain thermal comfort and additional financial stress felt by occupants to be able to pay for energy bills. There are a number of contributing factors resulting in fuel poverty, but poor quality and performing housing is a key contributor. Therefore, improving housing quality for all households is important to ensure that equity is delivered in relation to enhancing social value.

The social value for households is not just while they are living in their improved dwelling, but also when they come to sell (or rent) their dwelling. Research from around the world finds that there could be a sale premium of 15% (or more) resulting from improvements to design, quality, and performance [9, 20, 65–69]. In addition, research has found that houses with improved design, quality, and performance spent less time on the market, which is a significant benefit for the seller. Added sale/rent value comes not just from the dwelling itself but the quality of the local environment around the dwelling. Research has found that access to local amenities such as parks or having a view can add a further 15% (or more) to resale value [70–74]. Recent research from Sydney, Australia, even found that an increase of tree canopy on the street could increase sale values of property by AU\$33,000–AU\$61,000 [75].

Conversely, things such as a view of another apartment decreased resale value by up to 7% [70].

Aside from residential dwellings, the social value should also be considered in the other non-residential buildings (e.g. schools, offices). For example, CABE [20] found that in schools there could be an improvement of student learning scores of up to 26% due to improved design and quality outcomes (e.g. improved natural light) and that student drop-out rates could decrease by as much as 75% due to happier, healthier and more engaged students. In offices, productivity has been found to increase by up to 23% and there have been fewer sick days taken by staff. This could be translated to housing, especially important if working or education from home trends continue even after the COVID-19 pandemic recovery.

Beyond the building envelope, the area around a dwelling can significantly impact on social value including influencing the safety, security, quality, and performance of the dwelling, area, and overall urban environment. The space in between neighbouring buildings plays a critical role here as an opportunity to extend enhanced social value from individual buildings. Research from the United Kingdom [20, 76] found that safety and security is influenced by what happens immediately around the building. For example, making footpaths slightly wider and activating street frontages were found to reduce crime by 42%–90%. This then results in residents feeling safer in their own dwellings.

Additionally, increased tree coverage around dwellings can not only increase property value but critically helps reduce the heat island impact (whereby excess heat is trapped within the built environment) and can reduce ambient air temperatures during extreme weather days by 15°C or more [77–80]. This benefits individual dwellings by keeping them cooler to start with and reducing the need to use air conditioning while also reducing energy costs. Reducing energy demand during extreme weather events can have a tangible financial outcome through a reduction of costs for upgrading energy grids to cope with those peak weather events and help wider communities reduce issues, such as energy blackouts due to high demand.

Delivering this social value is sometimes easier said than done. In recent work exploring the idea and practices of ethical cities the authors [1] note that a number of issues throughout the built environment and the way key stakeholders engage, often with a lack of accountability, impacts the way that wider social value is able to be delivered across communities. The authors state that 'ethically oriented cities will ultimately be the ones that succeed in enhancing resilience, improving quality of life, creating productive economics and reducing the environmental burden for all residents' [1]. The authors note that ethical cities will address and enhance many of the points spoken above but require a number of changes across policy, the building industry and the way we understand and engage with social value and ethical considerations, if it is to be delivered.

There are a number of real-world case studies which are trying to deliver on the promise of enhanced social value. An early example was the BedZED development in the United Kingdom (see Case Study 9.1).

Case Study 9.1   Early exemplar of adding social value – BedZED.

BedZED in the United Kingdom was an early notable exemplar of sustainable housing development which demonstrated a range of improved social value outcomes for occupants, the surrounding community, and the wider construction industry [81].

Completed in 2002, the development contains 100 homes for approximately 220 residents and includes office space, a college, and community facilities. It has a mix of private market housing as well as homes for subsided rent and affordable home ownership, helping to address housing affordability issues. The design includes high levels of insulation, airtightness, and passive solar heating and solar photovoltaics, helping to significantly reduce energy consumption and utility costs [81, 82]. There is also a focus on reducing car usage through providing less on-site car parking and instead improvising cycling facilities and an on-site car-sharing club [83].

Analysis of the performance of the dwellings found that compared to similar standard developments in London there was a reduction of total carbon footprint (by 23%), electricity consumption (by 27%), CO_2 from home heating and cooling (32%), gas consumption (36%), water consumption (40%), and travel carbon footprint (53%) and a reduction in living costs of £1391 a year [83]. The building also had a significant focus on using local or reclaimed and recycled materials, helping to reduce its embodied energy environmental impact. Perhaps most significant was the focus on community and the way that the design facilitated the community coming together and, through the provision of open space and car-free streets, encouraged children to play outside more and neighbours to spend more time engaging with each other.

The benefits were more than just technical. The following is an example from one family who reflected on the benefits of living in BedZED, highlighting the wider social value provided:

'BedZED is different to anywhere else I have lived … to me it is very special and it's a shame that many more people don't have the opportunity to live in this sort of special place. Many former BedZEDders are still in touch with current residents due to the kind of people this place has brought together. Somehow the high density and closeness of residents has been reversed from a negative to a positive. It improves your quality of life to accept, consider and to an extent embrace those around you, reflecting that we share a small planet' [83].

The development and outcomes of BedZED inspired those involved to create the One Planet Living framework to guide other sustainable developments. Table 9.1 outlines the One Planet Living principles, which show that it is about more than just technology innovation and is more about delivering social value through a range of avenues.

Table 9.1 One Planet Living principles [83].

Health and happiness	Encouraging active, social, meaningful lives to promote good health and well-being
Equity and local economy	Creating safe, equitable places to live and work which support local prosperity and international fair trade
Culture and community	Nurturing local identity and heritage, empowering communities, and promoting a culture of sustainable living
Land and nature	Protecting and restoring land for the benefit of people and wildlife
Sustainable water	Using water efficiently, protecting local water resources, and reducing flooding and drought
Local and sustainable food	Promoting sustainable humane farming and healthy diets high in local, seasonal organic food, and vegetable protein
Travel and transport	Reducing the need to travel, encouraging walking, cycling, and low-carbon transport
Materials and products	Using materials from sustainable sources and promoting products which help people reduce consumption
Zero waste	Reducing consumption, reusing, and recycling to achieve zero waste and zero pollution
Zero carbon energy	Making buildings and manufacturing energy efficient and supplying all energy with renewables

Enhancing social value is not just occurring in new housing but also through retrofit of existing housing. Case Study 9.2 from the United Kingdom demonstrates the opportunities for delivering social value through deep retrofit to a Passivhaus (or Passive House) standard.

Case Study 9.2   Social housing Passive House refurbishment.

Enhancing social value is not just restricted to new dwellings and can be delivered in existing buildings through retrofit. Erneley Close is a social housing development in Manchester (United Kingdom) which underwent a refurbishment to EnerPHit Standard (the Passive House standard for refurbishments) in 2015. A post-occupancy evaluation undertaken on the refurbishment demonstrates the wide-ranging internal and external social value which can be created [3].

The project involved the owner (One Manchester) undertaking a refurbishment of 32 two-bedroom walk-up flats for a cost of £3.1 million. At the time of refurbishment, the flats were in disrepair. The goals for the project for the housing association were not only to reduce living costs and improve health and well-being outcomes for the occupants but to also create a new community greenspace and use the project as part of a wider regeneration of the area [3]. As David Power (Chief Executive, One Manchester) said, '… the reason why we've done this scheme is about creating long-term value for the neighbourhood and setting a standard for an area which needs wider regeneration' [84].

Research which monitored the performance of the refurbished dwellings found they performed significantly better than typical types of dwellings, with more stable indoor air temperatures and a reduction in the use of heating and cooling technologies [3]. This contributed to a reduction in energy costs for households, with savings of up to £100 a month reported by tenants. As one tenant reflected:

'Before all these works my flat was freezing. I was spending about £15 per week on heating the flat and even using fan heaters to get the temperature up. Since moving back in December, I've only used the heating once. It's really taken the pressure off, knowing we won't be spending an arm and a leg on keeping the house warm, day in, day out. More than that though, everyone here is just so proud of what's come out of this project – it's really put Erneley Close and Longsight on the map. There's a real community spirit here now … My little grandson calls the building 'Nanny's castle' because he says it's magical' [85].

As the quote above shows, the benefits were not just related to lower energy costs and a reduction in energy for heating and cooling. The post-occupancy research revealed that there was a significant uplift in community value and pride in the area – benefits which went beyond the individual dwelling [3]. Additionally, tenants reported that their health and well-being had improved. For example, several tenants spoke about having lower stress due to reduced energy bills. One tenant reported a reduction of asthma symptoms, and another stated their child was sleeping significantly better, which, in addition to the quietness in the dwelling due to improved building envelop, helped the child with their concentration while studying, potentially leading to better academic outcomes.

9.2 Delivering enhanced social value in the Australian residential context

This section presents how enhanced social value has manifested for households in different ways across a range of different housing types, tenures, and circumstances with three examples from our research in Australia. The first example draws upon our research during the COVID-19 pandemic, where it was found that stronger neighbourhood connections and relationships at a time of crisis could help increase the social value placed upon the residence. The second example is an environmentally sustainable new housing development in Cape Paterson (Victoria, Australia) that is delivering significantly improved housing performance and environmental outcomes, as well as housing which is more comfortable and affordable to live in. The final case is of a public housing development from regional Victoria, which delivered housing that is not only climate resilient but also improving the health and well-being of tenants.

9.2.1 Improved neighbourhood connection and relationships

In Chapter 7, we noted from our research of renters and landlords during the pandemic in Australia that there were a number of negative social value outcomes including perceptions over the security of housing, poor mental health, and stress over finances. However, our research also revealed a number of positive social value outcomes which emerged for some renters during the pandemic [86]. These positive social value outcomes that emerged during the COVID-19 pandemic could be considered in future built environment design and construction. These outcomes included that some tenants achieved greater social value through forming stronger relationships and connections with neighbours and the local area. In part, this was driven by the form and quality of rental dwellings. Many tenants explained that it was a challenge to spend substantial periods of time in rental dwellings, which often had issues around quality, performance, or design which were not well suited to living during a pandemic (see Chapter 7 for more on this). However, this did prompt many of these same tenants to look for any opportunity to safely and legally spend time outside the home (within the constraints of any lockdown restrictions). This often resulted in an uptake of physical exercise in nearby green spaces as a way to escape their dwellings and promote their mental health and well-being. For example, one renter stated:

> Mentally you don't like to be stuck inside … I think even just going out for some physical exercise has helped me.

Another notable improvement to social value for some tenants was around the development of improved relationships with neighbours. These tenants spoke of having ad-hoc relationships with their neighbours prior to COVID-19, but that the pandemic and associated governance restrictions and lockdowns meant they had got to know their neighbours much better. This included not just to have a conversation with, but that neighbours were helping each other out such as doing shopping for people, taking other neighbours dogs for walks, and just generally checking in on people's well-being:

> I talk to all of my neighbours, we are now all on a WhatsApp group, in our building, and we have been supporting one another through this whole lockdown business.

Similarly, another respondent stated:

> We have got really nice neighbours, I got much closer with them … [during the pandemic]. One of the neighbours works at a restaurant, and when they closed the restaurant he brought all the food home and shared it with everyone in the building. Another neighbour hurt his knee, so I have been walking his dog. I've been chatting to the neighbours a lot more, and the people I know in the building have become a lot closer.

While the COVID-19 pandemic highlighted a number of housing issues for people (see Chapter 7), there were some, perhaps surprising, enhanced social connections and value outcomes, as these examples above show. This highlights the importance of considering how to have greater opportunities to socially network with others within multi-occupancy buildings, both within building design and during occupancy.

9.2.2 Sustainable, comfortable, and affordable living

The Cape[1] is a sustainable housing development located about 1.5 hours southeast of Melbourne, Australia [87–89]. Once completed (the final land release occurred in 2022) the development will contain 230 new highly sustainable privately owned dwellings, as well as having a significant range of community amenities, such as around 50% open area and the inclusion of an urban farm. The development was driven by a need to demonstrate that new housing in Australia could be designed and delivered, which had a significantly reduced environmental impact, both in relation to the materials and technologies used during construction and across the life of the housing, but also in terms of reducing the energy consumption to live comfortably in the housing.

Through a collaborative design approach based upon evidence, it was determined that housing at The Cape would have to achieve a minimum 7.5 star thermal energy efficiency performance (on a scale of 0 = worst – 10 = best) with a requirement for other sustainability features such as a minimum 2.5 kilowatts of solar photovoltaics and 10,000 litres of rainwater storage. At 7.5 stars the energy required for heating and cooling (76 MJ/m^2/annum) is 40% less than compared to the regulated minimum at that time of 6.0 star (127 MJ/m^2/annum). This was in order to create housing which was not only had a low impact on the environment but delivered significant comfort, liveability, and financial benefits for households.

We have been evaluating the Cape from a range of perspectives [89]. So far, it is performing as expected. The improved performance of the dwellings resulted in a reduction in energy consumption of 88% compared to a typical new 6 star home of comparable size in Victoria. This translated to low energy bills, with households saving around AU\$2307/year and with many households in credit on their energy bills due to the production of onsite renewable energy. If these energy savings were replicated across all new housing in Australia across a decade it would lead to economic savings of more than AU\$5 billion. This would also help reduce the need for investment in the wider energy network by more than AU\$1 billion [90].

In relation to reduced environmental impacts, housing at the Cape is saving 4.4 tonnes of CO_2-e per year compared to a standard 6 star duel fuel newly built house in Victoria [89]. This increases to a saving of 11.2 tonnes of CO_2-e per year when the house is combined with an electric vehicle. This would be the equivalent of taking up to 80,000 cars off the Victorian roads each year.

For many households who participated in the above research, the financial and environmental benefits were secondary to the improved liveability and comfort of the dwellings and the improved community value delivered. For example, the participants reported how friendly the community was, even though it was still under development, and how the existing community welcomed new households with gift baskets. The community farm was also a significant social value addition, not only giving residents an opportunity to grow their own food but importantly contributing to the wider functioning of the development. The farm was so popular that some of the residents who were waiting for the home to be constructed and not yet living in the development would drive down from Melbourne to participate in some of the community gardening days.

It is also a community which is strongly focused on elevating those community relationships and connections. This starts with the design of the house, including not having a fence on the front of the property to encourage casual interactions amongst the community (e.g. when gardening the front lawn). Community members also share many things such as lawn movers or other equipment. The sharing does not just occur with the residents, but with the designers and builders involved in the construction; who share machinery, discuss how they help each other out, either when something might be going wrong with a build but also in terms of knowledge and learnings, thus reducing the need to 'reinvent the wheel' each time. The Cape development is demonstrating a range of social value being delivered at the household and community level.

9.2.3 Climate-resilient, improved health

The Department of Health and Human Services (a Victorian Government (Australia) department) recognised the need to deliver improved quality and performance of their public housing stock, not only to reduce environmental impact across government buildings but also as a way to deliver improved social value to public housing tenants. In order to understand how far beyond minimum building standards they should set proposed revisions to internal performance and quality requirements they funded the construction of four low-energy houses in regional Victoria, Australia. These were built to a 9 star standard (approaching passive house standard) and were built without any mechanical air conditioning. We led a longitudinal study that monitored these households and six control households in 'standard' department housing of a similar size and age [6, 91, 92].

From a technical perspective, the low-energy houses performed as expected, reducing electricity consumption by almost two-thirds. This resulted in the public housing tenants saving around AU$1000 per year in energy bills. While perhaps not a significant amount of money, for this low-income cohort it meant they had money to do things many people take for granted, such as buying fresh food, gifts for grandkids at Christmas, and even being able to go away on a short trip.

However, it was the improved thermal performance and climate resiliency as well as improvements to occupant health which were a significant value add. The assessment of the adaptive comfort criteria against the European thermal adaptive comfort standard, BS EN 15251, using monitored temperature and humidity data, shows that the low-energy houses were comfortable for 10% more of the time in summer for the living areas and 7% more of the time for the bedrooms compared with the control houses which even had air conditioning; this was all achieved without the use of additional air conditioning. The biggest benefit for thermal comfort was during extreme weather conditions such as heatwaves (with temperatures reaching upwards of 45°C during the study period), when the low-energy houses were significantly cooler than the control houses which were using air conditioning, reflecting the improved design and thermal performance of the dwellings. Figure 9.1 shows that on the second day of a heatwave, the best low-energy house was 16.6°C cooler compared to the worst control house (with air conditioning).

This improved comfort, particularly during the more extreme weather periods was something the low-energy residents spoke about during the interviews. For example, one resident in the low energy houses stated:

> Well we both feel the heat pretty well but when it was 42 degrees outside, it only got to 29 in here … when it was 3 degrees below zero this was 15 degrees inside on that morning, that's without any heaters being on, 15 degrees. So that's good.

This improved thermal performance of the low-energy houses was noticed by the occupants in relation to self-reported health improvements. For example, one occupant used to get pneumonia regularly during winter in their previous accommodation but had not had a case of it over the first three years in the

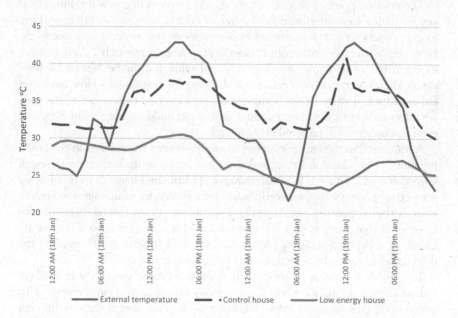

Figure 9.1 Temperature in living rooms of monitored houses and external temperature for 18–19 January 2013.

low-energy houses; an outcome they relate directly to the improved, and consistent, thermal performance. Their improvement of health was so significant that they were able to start working again and remove themselves from some welfare payments (which is a wider financial benefit for the government). Another occupant reported that they would get cramps in their legs when it got too cold, which made sleeping in winter difficult unless they were next to a heater. They were now able to sleep in their bed, which had a profound effect on their outlook on life. Again, this had seen a dramatic improvement in the low-energy house due to the improved thermal comfort.

9.3 Next steps: A policy push

This chapter starts from the premise that we can deliver enhanced social value in new and existing housing and that many of the issues discussed in the previous chapters could be avoided if housing quality and performance was improved in the design and construction process. While we are not naive to think that better attention during these phases (or during renovation/retrofit of existing housing) would eliminate all issues, it would certainly mitigate a significant number. In doing so, enhanced social value (or in some cases simply the avoidance of negative social value) would be achieved. This focus should not only be limited to the physical dwelling but the role that dwelling plays as part of the larger community and built environment and recognise that some social value will be enhanced by the connection of the dwelling to the area around the building.

The starting point of this is that we should be providing new dwellings (and any retrofit or renovation work) to be free of defects, with the ability to perform its core functions to the minimum expectations of the owner and/or occupant. However, as the avalanche of previous and emerging research clearly demonstrates, there is a significantly enhanced social value that can be delivered beyond simply the expectation of a defect-free dwelling for households. How to deliver enhanced social value in our housing must be addressed by changes to the construction industry (including building designers), building code regulators, and even through local or national planning schemes.

As discussed in the earlier chapters, there has been a culture within the building industry of taking shortcuts and there is often an attitude that 'near enough is good enough'. This has led to an increase of minor and major defects emerging across the housing sector, especially when only aiming for minimum standards to begin with. In countries like Australia, a significant reason this has been allowed to occur is that there is very low compliance checking, and even if defects are found, or a building industry professional is found to have contributed to the defect, there is often little recourse for the housing consumer.

To improve outcomes to begin with, there need to be much higher levels of independent compliance checks throughout the construction process. This would likely pick up issues before an owner takes possession of the dwelling but also subtly force building industry professionals to take more care and time when doing work the first time, if they knew there was a higher chance of unacceptable work being identified. Many sustainable housing rating tools/frameworks such as PassivHaus require compliance checks at certain points of the construction process to ensure that what is designed is delivered. This process is not just about identifying any potential issues from a sustainability or performance but can also pick up other quality issues.

Given that a dwelling is the most expensive thing most housing consumers will ever purchase, and given the long life of housing, it is not unreasonable to put in place additional checks and balances to ensure they are being delivered a product which at least meets minimum quality and safety requirements and that they are fit for purpose. While there may be an additional cost for (additional) compliance that would have to be covered (at least initially) by governments, it is likely the cost would be significantly less than dealing with minor and dangerous defects which emerge post-handover. These can run into the billions of dollars, as the example with flammable cladding demonstrates. The rate of compliance checking should be high enough that building industry stakeholders have the realistic knowledge that any of their work may be evaluated. Higher levels of compliance checks may be required, while the industry adjusts to new expectations of quality but may be able to be scaled back in the future, or for some building industry stakeholders, who continually demonstrate high-quality outcomes.

Policy makers also have a role to play in guiding, or requiring, the industry to go further with quality and performance requirements. Far too often the building industry in many locations around the world builds only to the minimum quality and performance requirements, which are about setting the lowest level acceptable, rather than what should be delivered [93, 94]. This has implications on performance and sustainability and particularly for the transition to a low carbon urban future. Given the long life of housing, any decision made about design and performance for new housing now will still likely be having an impact in 2050, where much of the wider climate and sustainability focus is targeting that we need to be at net zero carbon performance outcomes.

Lifting minimum regulations is one critical policy mechanism which should be leveraged. Minimum regulations have been critical for improving the quality and performance of housing around the world and are being used to drive performance towards zero carbon/energy performance outcomes [95, 96]. Jurisdictions like California and the European Union established medium-term pathways to make it clear to all stakeholders when various changes would be introduced or improved, for example, over a ten-year policy pathway. This helped to give stakeholders clarity on when things would change but also gave a clear direction of what was coming. Where not yet announced or planned, jurisdictions should create a clear policy pathway for when and how they will transition to low carbon/ sustainable housing and within this address issues of quality. Such quality and performance outcomes could be achievable within a ten-year period with the right mix of governance and innovation. This policy does not have to reinvent the wheel with several jurisdictions showing what does (and does not) work and a range of voluntary and mandatory quality and performance tools having been developed which can help lead the way.

To ensure the success of any changes to minimum quality and performance requirements, there must also be a focus on providing support to the building industry and professionals to ensure they are able to deliver the improved outcomes. This could be through professional development programs for those already in the industry or ensuring that initial training programs include more focus on these issues.

There are other options available as well which could drive both improve quality and performance such as offering to fast-track developments through planning approval processes if they are designed to significantly exceed minimum building code requirements and previous developments demonstrate high levels of quality (including being free of defects). This could also incorporate considerations such as what is offered around the building in relation to enhancing social value.

Housing consumers as well must do more to demand change. As the case studies explored in this chapter demonstrate, they will be the biggest winners from this enhanced social value. Housing consumers can help drive change by asking questions of developers, builders, or real estate agents about the quality and performance of any dwelling they are looking to buy or rent. Asking questions like what the thermal performance of the building envelope is, how much the house might cost to operate and maintain and what the opportunities are for disassembly and reuse/recycling at the end of life will not only give the housing consumer more information about their choices but also signal to the wider industry that there is a growing consumer cohort who want more from their housing. This should happen though in conjunction with regulatory and industry changes to shift thinking and practice around housing design, quality, performance, and sustainability. There are sufficient examples now around the world and across different dwelling types to show how we can do this, if the will and intent is there.

These changes would need to be framed with the consumer as the policy focus to ensure that outcomes support the enhancement of social value across the life of the dwelling.

9.4 Conclusions

The global evidence is clear – we have the capacity to be designing and constructing housing which can significantly enhance social value outcomes. Social value benefits for housing consumers include reduced environmental impact, reduced living costs, improved health and well-being, and the delivery of more resilient communities. There are benefits for the industry as well in relation to delivering higher-quality products and higher customer satisfaction, likely leading to less rework and less financial impact. Research also shows that those stakeholders who are leading the way can create a significant market advantage. For governments the benefits are wide reaching, including that if housing consumers are healthier, there will be less trips to doctors and hospitals and therefore help to ease the strain on many of these systems.

However, to deliver these benefits we must see a greater focus on delivering quality and performance through the design and construction phases (alongside consideration for end-of-life disassembly and reuse of materials). There is a multitude of examples of housing which deliver enhanced social value all around the world. However, delivering this type of housing at scale will require changes across the industry and by regulators to guide this change as well as housing consumers demanding better housing outcomes.

Note

1. Further information about this development is available at https://www.liveatthe cape.com.au/

References

1. Barrett, B., R. Horne, and J. Fien, *Ethical cities.* 2021, Routledge.
2. Willand, N., C. Maller, and I. Ridley, *Addressing health and equity in residential low carbon transitions – Insights from a pragmatic retrofit evaluation in Australia.* Energy Research & Social Science, 2019. **53**: pp. 68–84. DOI: https://doi.org/10.1016/j. erss.2019.02.017
3. Sherriff, G., P. Martin, and B. Roberts, *Erneley Close passive house retrofit: Resident experiences and building performance in retrofit to passive house standard.* 2018, University of Salford. http://usir.salford.ac.uk/46328/
4. Nelson, A., *Small is necessary: Shared living on a shared planet.* 2018, Pluto Press.
5. Horne, R., *Housing sustainability in low carbon cities.* 2018, London: Taylor & Francis Ltd.
6. Moore, T., I. Ridley, Y. Strengers, C. Maller, and R. Horne, *Dwelling performance and adaptive summer comfort in low-income Australian households.* Building Research & Information, 2017: pp. 1–14. DOI: 10.1080/09613218.2016.1139906.
7. Moore, T., L. Nicholls, Y. Strengers, C. Maller, and R. Horne, *Benefits and challenges of energy efficient social housing.* Energy Procedia, 2017. **121**: pp. 300–307. DOI: https://doi.org/10.1016/j.egypro.2017.08.031
8. Montgomery, C., *Happy city: Transforming our lives through urban design.* 2013, Penguin UK.
9. Yudelson, J., *The green building revolution.* 2010, Island Press.
10. Sherriff, G., T. Moore, S. Berry, A. Ambrose, B. Goodchild, and A. Maye-Banbury, *Coping with extremes, creating comfort: User experiences of 'low-energy' homes in Australia.* Energy Research & Social Science, 2019. **51**: pp. 44–54. DOI: https://doi. org/10.1016/j.erss.2018.12.008
11. Pears, A., and T. Moore, *Decarbonising household energy use: The smart meter revolution and beyond.* In: *Decarbonising the built environment,* 2019. Springer. pp. 99–115.
12. Moore, T., and A. Doyon, *The uncommon nightingale: Sustainable housing innovation in Australia.* Sustainability, 2018. **10**(10): p. 3469.
13. Moore, T., F. de Haan, R. Horne, and B. Gleeson, *Urban sustainability transitions. Australian Cases – International perspectives.* Theory and Practice of Urban Sustainability Transitions. 2018, Singapore: Springer.
14. IPCC, *Fifth assessment report. Climate Change 2014: Working group III: Mitigation of climate change.* 2014, Valencia, Spain: Intergovernmental Panel on Climate Change.
15. Berry, S., and K. Davidson, *Zero energy homes – Are they economically viable?* Energy Policy, 2015. **85**: pp. 12–21. DOI: https://doi.org/10.1016/j.enpol.2015.05.009
16. Berry, S., D. Whaley, K. Davidson, and W. Saman, *Do the numbers stack up? Lessons from a zero carbon housing estate.* Renewable Energy, 2014. **67**: pp. 80–89. DOI: 10.1016/j.renene.2013.11.031.

17. Berry, S., *The technical and economic feasibility of applying a net zero carbon standard for new housing*. In: *School of engineering, division of information technology, engineering and the environment*. 2014, University of South Australia: Adelaide.

18. Mithraratne, N., B. Vale, and R. Vale, *Sustainable living: The role of whole life costs and values*. 2007, Oxford: Elsevier Limited.

19. Berry, S., D. Whaley, K. Davidson, and W. Saman, *Near zero energy homes – What do users think?* Energy Policy, 2014. **73**: pp. 127–137. DOI: 10.1016/j.enpol.2014.05.011.

20. CABE, *The value of good design. How buildings and spaces create economic and social value*. 2002, London: Commission for Architecture and the Built Environment. https://www.designcouncil.org.uk/sites/default/files/asset/document/the-value-of-good-design.pdf

21. Daniel, L., E. Baker, and T. Williamson, *Cold housing in mild-climate countries: A study of indoor environmental quality and comfort preferences in homes, Adelaide, Australia*. Building and Environment, 2019. **151**: pp. 207–218. DOI: 10.1016/j.buildenv.2019.01.037.

22. Pevalin, D.J., M.P. Taylor, and J. Todd, *The dynamics of unhealthy housing in the UK: A panel data analysis*. Housing Studies, 2008. **23**(5): pp. 679–695. DOI: 10.1080/02673030802253848.

23. Pevalin, D.J., A. Reeves, E. Baker, and R. Bentley, *The impact of persistent poor housing conditions on mental health: A longitudinal population-based study*. Preventive Medicine, 2017. **105**: pp. 304–310. DOI: 10.1016/j.ypmed.2017.09.020.

24. Baker, E., A. Beer, L. Lester, D. Pevalin, C. Whitehead, and R. Bentley, *Is housing a health insult?* International Journal of Environmental Research and Public Health, 2017. **14**(6). DOI: 10.3390/ijerph14060567.

25. Raynor, K., I. Wiesel, and B. Bentley, *Why staying home during a pandemic can increase risk for some*, in *Discussion Paper*. 2020, The University of Melbourne: Affordable Housing Hallmark Research Initiative. 06/07/2020. https://msd.unimelb.edu.au/atrium/why-staying-home-during-a-pandemic-can-increase-risk-for-some#:~:text=Confronted%20with%20high%20levels%20of,with%20higher%20risks%20of%20contagion.&text=Homeowners%20face%20other%20restrictions%20in%20their%20ability%20to%20stay%20home

26. Farha, L., *COVID-19 Guidance Note: Protecting renters and mortgage payers*, 2020. United Nations [Online]. 07/08/2020 https://www.ohchr.org/Documents/Issues/Housing/SR_housing_COVID-19_guidance_rent_and_mortgage_payers.pdf

27. Rogers, D., and E. Power, *Housing policy and The COVID-19 pandemic: The importance of housing research during this health emergency*. International Journal of Housing Policy, 2020. **20**(2): pp. 177–183. DOI: 10.1080/19491247.2020.1756599.

28. AHURI, *COVID-19 mortgage stress creating uncertain housing futures. Pandemic recession leaves households in mortgage stress and socio-economic hardship*, 2020. https://www.ahuri.edu.au/policy/ahuri-briefs/covid-19-mortgage-stress-creating-uncertain-housing-futures

29. Bentley, R., and E. Baker, *Housing at the frontline of the COVID-19 challenge: A commentary on 'rising home values and COVID-19 case rates in Massachusetts'*. Social Science & Medicine, 2020. **265**: p. 113534. DOI: 10.1016/j.socscimed.2020.113534.

30. Gualano, M.R., G. Lo Moro, G. Voglino, F. Bert, and R. Siliquini, *Effects of COVID-19 lockdown on mental health and sleep disturbances in Italy*. International Journal of Environmental Research and Public Health, 2020. **17**(13): p. 4779. DOI: 10.3390/ijerph17134779.

31. Manville, M., P. Monkkonen, and M. Lens, *COVID-19 and renter distress: Evidence from Los Angeles. August 2020*. 2020, UCLA. https://escholarship.org/content/qt7sv4n7pr/qt7sv4n7pr.pdf

32. Rosenberg, A., D.E. Keene, P. Schlesinger, A.K. Groves, and K.M. Blankenship, *COVID-19 and hidden housing vulnerabilities: Implications for health equity, new haven, Connecticut*. AIDS and Behavior, 2020. **24**(7): pp. 2007–2008. DOI: 10.1007/s10461-020-02921-2.

33. Rossi, R., V. Socci, D. Talevi, S. Mensi, C. Niolu, F. Pacitti, A.D. Marco, A. Rossi, A. Siracusano, and G. Di Lorenzo, *COVID-19 pandemic and lockdown measures impact on mental health among the general population in Italy.* Frontiers in Psychiatry, 2020. **11**(790). DOI: 10.3389/fpsyt.2020.00790.

34. Ambrose, A., W. Baker, G. Sherriff, and J. Chambers, *Cold comfort: Covid-19, lockdown and the coping strategies of fuel poor households.* Energy Reports, 2021. DOI: 10.1016/j.egyr.2021.08.175.

35. Bower, M., C. Buckle, E. Rugel, A. Donohoe-Bales, L. McGrath, K. Gournay, E. Barrett, P. Phibbs, and M. Teesson, *'Trapped', 'anxious' and 'traumatised': COVID-19 intensified the impact of housing inequality on Australians' mental health.* International Journal of Housing Policy, 2021. pp. 1–32. DOI: 10.1080/19491247.2021.1940686.

36. Gurney, C.M., *Dangerous liaisons? Applying the social harm perspective to the social inequality, housing and health trifecta during the Covid-19 pandemic.* International Journal of Housing Policy, 2021. pp. 1–28. DOI: 10.1080/19491247.2021.1971033.

37. The Lancet, *Redefining vulnerability in the era of COVID-19.* The Lancet, 2020. **395**(10230): p. 1089. DOI: 10.1016/s0140-6736(20)30757-1.

38. Yancy, C., *Covid-19 and African Americans.* JAMA Network, 2020. **323**(19): pp. 1891–1892.

39. Horne, R., N. Willand, L. Dorignon, and B. Middha, *Housing inequalities and resilience: The lived experience of COVID-19.* International Journal of Housing Policy, 2021. pp. 1–25. DOI: 10.1080/19491247.2021.2002659.

40. Chapman, R., P. Howden-Chapman, H. Viggers, D. O'Dea, and M. Kennedy, *Retrofitting houses with insulation: A cost-benefit analysis of a randomised community trial.* Journal of Epidemiology and Community Health, 2009. **63**(4): pp. 271–277. DOI: 10.1136/jech.2007.070037.

41. Huang, C., A. Barnett, Z. Xu, C. Chu, X. Wang, L. Turner, and S. Tong, *Managing the health effects of temperature in response to climate change: Challenges ahead.* Environmental Health Perspectives 2013. **121**: pp. 415–419. DOI: 10.1289/ehp.1206025.

42. Kosatsky, T., *The 2003 European heat waves.* Eurosurveillance, 2005. **10**(7): pp. 148–149.

43. Porto Valente, C., A. Morris, and S.J. Wilkinson, *Energy poverty, housing and health: The lived experience of older low-income Australians.* Building Research & Information, 2021. pp. 1–13. DOI: 10.1080/09613218.2021.1968293.

44. Martin, W., *Exploring the mental health impact on private flat owners in residential buildings with external combustible cladding.* BJPsych Open, 2021. **7**(S1): pp. S268–S269. DOI: 10.1192/bjo.2021.715.

45. Gower, A., *Energy justice in apartment buildings and the spatial scale of energy sustainable design regulations in Australia and the UK.* Frontiers in Sustainable Cities, 2021. **3**(27). DOI: 10.3389/frsc.2021.644418.

46. Daly, D., T. Harada, M. Tibbs, P. Cooper, G. Waitt, and F. Tartarini, *Indoor temperatures and energy use in NSW social housing.* Energy and Buildings, 2021. **249**: p. 111240. DOI: 10.1016/j.enbuild.2021.111240.

47. Heffernan, T., E. Heffernan, N. Reynolds, W.J. Lee, and P. Cooper, *Towards an environmentally sustainable rental housing sector.* Housing Studies, 2020. pp. 1–24 DOI: 10.1080/02673037.2019.1709626.

48. Foster, S., P. Hooper, A. Kleeman, E. Martino, and B. Giles-Corti, *The high life: A policy audit of apartment design guidelines and their potential to promote residents' health and wellbeing.* Cities, 2020. **96**: p. 102420. DOI: 10.1016/j.cities.2019.102420.

49. Daniel, L., T. Moore, E. Baker, A. Beer, N. Willand, R. Horne, and C. Hamilton, *Warm, cool and energy-affordable housing solutions for low-income renters, AHURI final report no. 338*, 2020. Melbourne: Australian Housing and Urban Research Institute Limited. https://www.ahuri.edu.au/research/final-reports/338; DOI: 10.18408/ahuri-3122801.

50. Baker, E., N.T.A. Pham, L. Daniel, and R. Bentley, *New evidence on mental health and housing affordability in cities: A quantile regression approach.* Cities, 2020. **96:** p. 102455. DOI: 10.1016/j.cities.2019.102455.

51. Baker, E., L.H. Lester, R. Bentley, and A. Beer, *Poor housing quality: Prevalence and health effects.* Journal of Prevention & Intervention in the Community, 2016. **44**(4): pp. 219–232. DOI: 10.1080/10852352.2016.1197714.

52. Garrett, H., M. Mackay, S. Nicol, J. Piddington, and M. Roys, *The cost of poor housing in England. 2021 Briefing paper,* 2021. London: BRE. https://files.bregroup.com/research/BRE_Report_the_cost_of_poor_housing_2021.pdf

53. Colton, M.D., P. MacNaughton, J. Vallarino, J. Kane, M. Bennett-Fripp, J.D. Spengler, and G. Adamkiewicz, *Indoor air quality in green vs conventional multifamily low-income housing.* Environmental Science & Technology, 2014. **48**(14): pp. 7833–7841. DOI: 10.1021/es501489u.

54. Poor, J.A., D. Thorpe, and Y. Goh, *The key-components of sustainable housing design for Australian small size housing.* International Journal of GEOMATE, 2018. **15**(49): pp. 23–29.

55. Strengers, Y., and L. Nicholls, *Convenience and energy consumption in the smart home of the future: Industry visions from Australia and beyond.* Energy Research & Social Science, 2017. **32:** pp. 86–93. DOI: 10.1016/j.erss.2017.02.008.

56. Maalsen, S., *Revising the smart home as assemblage.* Housing Studies, 2020. **35**(9): pp. 1534–1549. DOI: 10.1080/02673037.2019.1655531.

57. Moore, T., *Facilitating a transition to zero emission new housing in Australia: Costs, benefits and direction for policy.* In: *School of global, urban and social studies,* 2012. Melbourne: RMIT University.

58. Matthew, P., and P. Leardini, *Towards net zero energy for older apartment buildings in Brisbane.* Energy Procedia, 2017. **121:** pp. 3–10.

59. Sovacool, B., M. Lipson, and R. Chard, *Temporality, vulnerability, and energy justice in household low carbon innovations.* Energy Policy, 2019. **128:** pp. 495–504. DOI: 10.1016/j.enpol.2019.01.010.

60. Willand, N., and R. Horne, *'They are grinding us into the ground' – The lived experience of (in)energy justice amongst low-income older households.* Applied Energy, 2018. **226:** pp. 61–70 DOI: 10.1016/j.apenergy.2018.05.079.

61. Bouzarovski, S., *Energy poverty: (Dis)Assembling Europe's infrastructural divide.* 2018, Springer Nature. DOI: 10.1007/978-3-319-69299-9.

62. O'Sullivan, K.C., P. Howden-Chapman, D. Sim, J. Stanley, R.L. Rowan, I.K. Harris Clark, and L. Morrison, *Cool? Young people investigate living in cold housing and fuel poverty. A mixed methods action research study.* SSM – Population Health, 2017. **3:** pp. 66–74. DOI: 10.1016/j.ssmph.2016.12.006.

63. Boardman, B., *Fuel poverty,* In *International encyclopedia of housing and home,* S.J. Smith, ed. 2012, Elsevier: San Diego. p. 221–225 DOI: http://dx.doi.org/10.1016/B978-0-08-047163-1.00552-X

64. Walker, G., and R. Day, *Fuel poverty as injustice: Integrating distribution, recognition and procedure in the struggle for affordable warmth.* Energy Policy, 2012. **49**(0): pp. 69–75. DOI: 10.1016/j.enpol.2012.01.044.

65. DEWHA, *Energy efficiency rating and house price in the ACT.* 2008, Canberra: Department of the Environment Water Heritage and the Arts.

66. Nevin, R., and G. Watson, *Evidence of rational market valuations for home energy efficiency.* Appraisal Journal, 1998. **66:** pp. 401–409.

67. Bloom, B., M.C. Nobe, and M.D. Nobe, *Valuing green home designs: A study of ENERGY STAR homes.* Journal of Sustainable Real Estate, 2011. **3**(1): pp. 109–126.

68. Hoen, B., R. Wiser, P. Cappers, and M. Thayer, *An analysis of the effects of residential photovoltaic energy systems on home sales prices in California.* 2011, San Diego: Ernest Orlando Lawrence Berkeley National Laboratory.

69. Kok, N., and M. Kahn, *The value of green labels in the California housing market.* 2012, California: UCLA Institute of the Environment and Sustainability Los Angeles, July 2012.

70. CABE, *The value of public space. How high quality parks and public spaces create economic, social and environmental value.* 2003, London: Commission for Architecture and the Built Environment.

71. Kong, F., H. Yin, and N. Nakagoshi, *Using GIS and landscape metrics in the hedonic price modeling of the amenity value of urban green space: A case study in Jinan City, China.* Landscape and Urban Planning, 2007. **79**(3–4): pp. 240–252. DOI: 10.1016/j.landurbplan.2006.02.013.

72. Sander, H.A., and S. Polasky, *The value of views and open space: Estimates from a hedonic pricing model for Ramsey County, Minnesota, USA.* Land Use Policy, 2009. **26**(3): pp. 837–845. DOI: 10.1016/j.landusepol.2008.10.009.

73. Tyrväinen, L., and A. Miettinen, *Property prices and urban forest amenities.* Journal of Environmental Economics and Management, 2000. **39**(2): pp. 205–223. DOI: 10.1006/jeem.1999.1097.

74. Bourassa, S., M. Hoesli, and J. Sun, *What's in a view?* Environment and Planning A, 2004. **36**: pp. 1427–1450.

75. Swinbourne, R., and J. Rosenwax, *Green infrastructure: A vital step to brilliant Australian cities.* 2017, Technical report, AECOM.

76. CABE, *Paved with gold. The real value of good street design.* 2007, London: Commission for Architecture and the Built Environment.

77. Santamouris, M., L. Ding, and P. Osmond, *Urban heat island mitigation.* In: *Decarbonising the built environment,* 2019. Springer. pp. 337–355.

78. Arifwidodo, S.D., P. Ratanawichit, and O. Chandrasiri. *Understanding the implications of urban heat island effects on household energy consumption and public health in Southeast Asian cities: Evidence from Thailand and Indonesia.* In AUC 2019, 2021. Singapore: Springer Singapore.

79. Roxon, J., F.J. Ulm, and R.J.M. Pellenq, *Urban heat island impact on state residential energy cost and CO_2 emissions in the United States.* Urban Climate, 2020. **31**: p. 100546. DOI: 10.1016/j.uclim.2019.100546.

80. Duncan, J.M.A., B. Boruff, A. Saunders, Q. Sun, J. Hurley, and M. Amati, *Turning down the heat: An enhanced understanding of the relationship between urban vegetation and surface temperature at the city scale.* Science of the Total Environment, 2019. **656**: pp. 118–128 DOI: 10.1016/j.scitotenv.2018.11.223.

81. Chance, T., *Towards sustainable residential communities: The Beddington Zero Energy Development (BedZED) and beyond.* Environment and Urbanization, 2009. **21**(2): pp. 527–544. DOI: 10.1177/0956247809339007.

82. Hodge, J., and J. Haltrecht, *BedZED seven years on. The impact of the UK's best known eco-village and its residents.* 2009, London: BioRegional.

83. Schoon, N., *The BedZED story: The UK's first large-scale, mixed-use eco-village.* 2016, Wallington: BioRegional. 11/29/2021 https://www.bioregional.com

84. PassivHaus Trust, *UK Passivhaus Awards 2015: Large projects category – Erneley Close,* 2015. London: YouTube. https://www.youtube.com/watch?v=SEr079CtPeE&ab_channel=PassivhausTrust

85. PassivHaus Trust. *UK PassivHaus Awards 2015. Erneley Close Retrofit,* 2015. https://www.passivhaustrust.org.uk/UserFiles/File/UK%20PH%20Awards/2015/2015%20posters/ERNELEY%20CLOSE_Poster%20web(1).pdf

86. Oswald, D., T. Moore, and E. Baker, *Post pandemic landlord-renter relationships in Australia, AHURI final report no. 344,* 2020. Melbourne: Australian Housing and Urban Research Institute Limited. https://www.ahuri.edu.au/research/final-reports/344; DOI: 10.18408/ahuri5325901.

87. The Cape, *Welcome to the Cape,* 2021. https://www.liveatthecape.com.au/

88. Szatow, A., *Cape Paterson Ecovillage: Zero carbon study peer review*, 2011. Melbourne: Prepared for the Cape Paterson Partnership and Sustainability Victoria.

89. Moore, T., N. Willand, S. Holdsworth, S. Berry, D. Whaley, G. Sheriff, A. Ambrose, and L. Dixon, *Evaluating the cape: pre and post occupancy evaluation update January 2020*, 2020. Melbourne: RMIT University and Renew. https://renew.org.au/wp-content/uploads/2020/01/Evaluating-The-Cape-research-RMIT_Renew-January-2020.pdf

90. ASBEC, ClimateWorks, *The Bottom line the household impacts of delaying improved energy requirements in the building code*, 2018. Sydney: Australian Sustainable Built Environment Council. https://apo.org.au/node/131876

91. Moore, T., Y. Strengers, and C. Maller, *Utilising mixed methods research to inform low-carbon social housing performance policy*. Urban Policy and Research, 2016. **34**(3): pp. 240–255. DOI: 10.1080/08111146.2015.1077805.

92. Moore, T., Y. Strengers, C. Maller, I. Ridley, L. Nicholls, and R. Horne, *Horsham catalyst research and evaluation. Final report*, 2016. Melbourne: RMIT University.

93. Berry, S., T. Moore, and M. Ambrose, *Flexibility versus certainty: The experience of mandating a building sustainability index to deliver thermally comfortable homes*. Energy Policy, 2019. **133**: p. 110926. DOI: 10.1016/j.enpol.2019.110926.

94. Moore, T., S. Berry, and M. Ambrose, *Aiming for mediocrity: The case of Australian housing thermal performance*. Energy Policy, 2019. **132**: pp. 602–610. DOI: 10.1016/j.enpol.2019.06.017.

95. Moore, T., *Modelling the through-life costs and benefits of detached zero (net) energy housing in Melbourne, Australia*. Energy and Buildings, 2014. **70**(0): pp. 463–471. DOI: 10.1016/j.enbuild.2013.11.084.

96. Doyon, A., and T. Moore, *The role of mandatory and voluntary approaches for a sustainable housing transition: Evidence from Vancouver and Melbourne*. Urban Policy and Research, 2020. pp. 1–17. DOI: 10.1080/08111146.2020.1768841.

10 Conclusions

10.1 The need for change

As a society we expect our built environment to meet certain quality and performance outcomes. However, over recent decades, there have been increasing instances where building quality and performance are falling short of not only consumer expectations but also regulated requirements, especially in the residential sector.

The recent emergence of building cracks and combustible cladding are just two notable examples of dangerous building defects. Such defects have meant dwellings are unsafe, unsellable, and under-performing, resulting in a complete loss of financial and social value for households. These building quality issues are too often left to the consumer to deal with, meaning costly rectification work to make their dwellings safe and functional again. The consumer is an essential stakeholder within the built environment, yet the attention they deserve in research and practice is often overlooked.

In this book we have explored the issues of housing quality and performance from the perspective of the housing consumer and the implications on social value. This focus has often been limited in policy development, research, and industry practice. We have defined social value as: the social impact that is created by organisations and key stakeholders within the built environment for the lives of those affected by their activities [1].

When social value is created within the built environment, it can have a range of implications for housing consumers, including improved health and well-being, financial outcomes, and relationships with neighbours [2–6]. Social value can also be reduced through, for example, the emergence of residential defects. When the defects are minor, there can be a reduced appreciation for the home, and when they are major, there can be safety, health, and well-being issues.

We have stressed throughout this book that we should not only be trying to reduce negative social value impacts in design and construction processes but use these to significantly enhance social value outcomes. This is achieved by going beyond 'minimum' and 'fit for purpose' designs and creating socially sensitive infrastructure or architecture that positively contributes to individuals, households and communities, not only for the present but into the future.

DOI: 10.1201/9781003176336-10

10.2 Key takeaways

Drawing upon international case studies and our own research [2, 7–12] we have highlighted key takeaways within each chapter of the book (aside from the introduction).

In Chapter 2, we highlight how there has long been a focus on the physical building components and issues that arise during design and construction, but with little focus on the human implications have from defects post-handover. In addition, there have been a range of dangerous defects which have emerged, which we argue requires a new definition to better capture the size, scale, and impact of defects. These dangerous defects, such as combustible cladding and asbestos, are in a league of their own. The definition for a 'dangerous defect' that we propose is:

> A major shortfall in the building performance that emerges after the build-ing is in use, which is not rectified in a timely manner, is costly to address, and poses a continuous risk to the occupants' safety, health or well-being, that can last for years.

With a definition of dangerous defects, there is a starting point for support and focus. For example, there could be specific government financial and well-being packages for defects that fit within the definition of 'dangerous defect'. This would help separate from the defects that would not qualify from this support (e.g. a rattling window) and defects that would. Dangerous defects need this separation, as they are not 'temporary' problems that are easily fixed, and they can cause sig-nificant reductions in safety, health, well-being, and social value.

In Chapters 3–6, we use the example of flammable cladding as a dangerous defect to highlight the various well-being, financial and response challenges that consumers face. Our research shows that the costs associated with defect rectification go beyond just the financial cost of the physical building materi-als. For example, there can be financial costs for legal fees, improving short-term safety, engaging industry experts, and increased insurance premiums. Typically, these other costs fall on the consumer and outside of any government or industry support.

Additionally, households in impacted buildings also demonstrate a range of negative health and well-being outcomes such as increased stress, depression and other emotional issues resulting from trying to deal with having building defects rectified. There is often little support provided for occupant health and well-being resulting from defects.

A number of these points are exacerbated in multi-occupancy dwellings where consensus decision-making is often hard to achieve on how to respond. While professional strata managers and owners corporations have a key role to play to address these shortcomings, the evidence finds that in relation to larger defects these governance structures have a lack of capacity and skills to be able to provide the required support.

When a dangerous defect emerges, such as flammable cladding, there are often significant financial costs, negative well-being implications, and challenges with other residents in multiple occupancy buildings. In these situations, there can be a complete loss of any social value placed upon their individual units and the building as a whole. This can often be reflected in the financial value of the property, with some apartments with flammable cladding in the United Kingdom being deemed worth £0.

In Chapter 7, we discuss how the rental market also presents additional challenges for both landlords and tenants. Private rental housing for example is often already amongst the poorest housing quality. Tensions can arise between tenants and landlords when requesting defects or performance issues to be addressed. The COVID-19 crisis has amplified problems within the private rental sector in many regions of the world, including tenants' reluctance to report defects due to concerns of housing insecurity and refusal and delays for tenants to have defects fixed. This has negatively impacted tenant social value and highlighted a lack of clear and independent frameworks to address these issues and protect tenants.

In Chapter 8, we emphasise how governments and industry are not prepared to deal with dangerous defects, such as flammable cladding. This means that any response is reactive and often made up 'on the run', leaving housing consumers in the dark and struggling to know what to do. This can add significant time to getting defects fixed.

It is also clear that there is a lack of consumer protection when defects (of any scale) emerge. While many jurisdictions have things like building warranties, these have been found to be limited in what they cover and often push the cost of any rectification back onto the housing consumer. These warranties and other consumer protection have also been found to be inadequate when dealing with dangerous defects.

Where industry stakeholders are not able to take responsibility for their failings, some governments have stepped in with some financial and other support to address dangerous defects. While this has helped, very rarely have governments contributed 100% of the rectification costs and even more rarely contributed to wider costs incurred by impacted households.

We also highlight how the building regulations that have been developed in recent decades have largely been centred around trying to make the construction industry 'work', with limited consideration of robust housing consumer protection. There is a need to reframe policy development to place the housing consumer at the heart of policy making. This would complement current industry-based policy approaches and provide greater support and social value for consumers that encounter dangerous residential defects. This would not only help improve outcomes when a significant defect issue emerges but reduce the likelihood of them occurring to begin with.

In Chapter 9, we emphasise that the focus should not only be about addressing defects when they arise but proactively attempting to stop them from manifesting in the first place, through high-quality design and construction. We demonstrate that it is possible to create high-quality and sustainable housing that significantly enhances social value for consumers through research and international case studies. These examples highlight how households have been found to be happier, healthier, have lower bills, stronger community connection, and a range of other positive social value outcomes. This can be done with the materials, technology, and knowledge that are currently available.

10.3 Next steps

Overall, there needs a fundamental re-think into how to shape the built world for a better future. We need to consider how we can improve outcomes for consumers so they can live in homes that are not only defect free but are sustainable and comfortable, with socially valuable surroundings both within the home and the local neighbourhood. There should be a requirement to create more homes that consumers can be proud of. To take steps towards this goal, there needs to be a substantial shift in thinking and approaches from key stakeholders in government and industry.

The government should consider ways to improve consumer protection within the built environment. In the current consumer protection systems, it is often too simple for industry stakeholders to ignore returning to fix defects that emerge in the occupancy stages (despite warranty periods). The ways of regulating the construction industry clearly need to be improved, since when dangerous defects arise, such as flammable cladding, asbestos, or widespread leaky buildings, the lack of consumer protection across thousands of consumers is clearly exposed. A more robust framework for dealing with such defects is required from the government including:

- having adequate disaster relief available for these types of human-made issues,
- more sophisticated ways of identifying all the buildings affected (e.g. building material passports),
- warranties that are more robust that have a greater emphasis on a timely fix,
- affordable consumer access the legal representatives and action,
- appropriate financial and mental health support for affected consumers,
- increased government resources to help advise and support affected consumers, and
- appropriate rehousing and other living support, where necessary.

Beyond improving this response to poor building standards, there is a need to have a much greater emphasis on designing and construction high-quality buildings in the first place and subsequently in retrofits. Government policy that places the consumer at the centre will consider driving these high-quality standards through policy that encourages, for example, improved thermal performance of dwellings which will not only reduce energy consumption but improve occupant health and well-being. We note a number of examples from around the world where this has, or is, occurring, such as BedZED in the United Kingdom and the Cape in Australia.

The industry also needs to be part of the solution for driving change. This would represent a shift from an industry that is largely focused production, time, cost, and price to considering other project goals more closely, including quality, sustainability, and demonstrating corporate social responsibility. There are a minority of companies that have led the way with this shift within the built environment. For example, there are companies recognised as B-Corps, who have demonstrated they do not simply focus on profits and satisfying shareholders financially but have proven to create social value through the work they do for the consumer and wider society. It is also worth noting that the housing consumer themselves have a role to play in pursuing a better-built environment. This is primarily through creating the pressure and demand for high-quality housing that is sustainable and socially valuable, rather than settling for low-cost and poor-quality options.

A key challenge to addressing building quality and defect issues is the way researchers, policy makers, and the wider construction industry have typically considered the issues. It is recommended that in the short term, there may be an initial focus on protecting social value in housing through mitigating the likeliness of minor and major defects. In the medium-term, there could be a focus on creating widespread high-quality buildings that consumers can be proud of. In the longer term, it is important to focus on enhancing social value for all consumers, including those most vulnerable in our communities (e.g. low-income households) to ensure an equitable transition to improve building quality and performance.

References

1. Raiden, A., M. Loosemore, A. King, and C. Gorse, *Social value in construction*. 2019, Oxon: Routledge.
2. Oswald, D., *Homeowner vulnerability in residential buildings with flammable cladding.* Safety Science, 2021. **136**. DOI: https://doi.org/10.1016/j.ssci.2021.105185.
3. Moore, T., Y. Strengers, C. Maller, I. Ridley, L. Nicholls, and R. Horne, *Horsham catalyst research and evaluation*. 2016, Melbourne: RMIT University.
4. Horne, R., *Housing sustainability in low carbon cities*. 2018, London: Taylor & Francis Ltd.
5. Montgomery, C., *Happy city: Transforming our lives through urban design.* 2013, London: Penguin UK.
6. CABE, *The value of good design. How buildings and spaces create economic and social value*, 2002. London: Commission for Architecture and the Built Environment. https://www.designcouncil.org.uk/sites/default/files/asset/document/the-value-of-good-design.pdf
7. Oswald, D., T. Moore, and E. Baker, *Post pandemic landlord-renter relationships in Australia, AHURI Final Report No. 344*, 2020. Melbourne: Australian Housing and Urban Research Institute Limited. https://www.ahuri.edu.au/research/final-reports/344; DOI: 10.18408/ahuri5325901.
8. Oswald, D., T. Moore, and S. Lockrey, *Flammable cladding and the effects on homeowner well-being.* Housing Studies, 2021: pp. 1–20. DOI: 10.1080/02673037.2021.1887458.

9. Oswald, D., T. Moore, and S. Lockrey, *Combustible costs! Financial implications of flammable cladding for homeowners*. International Journal of Housing Policy, 2021: pp. 1–21. DOI: 10.1080/19491247.2021.1893119.

10. Moore, T., N. Willand, S. Holdsworth, S. Berry, D. Whaley, G. Sheriff, A. Ambrose, and L. Dixon, *Evaluating the cape: Pre and post occupancy evaluation update January 2020*, 2020. Melbourne: RMIT University and Renew. https://renew.org.au/wp-content/uploads/2020/01/Evaluating-The-Cape-research-RMIT_Renew-January-2020.pdf

11. Moore, T., Y. Strengers, C. Maller, I. Ridley, L. Nicholls, and R. Horne, *Horsham Catalyst research and evaluation. Final Report*, 2016. Melbourne: RMIT University.

12. Oswald, D., T. Moore, and E. Baker, *Exploring the well-being of renters during the COVID-19 pandemic*. International Journal of Housing Policy, 2022: pp. 1–21. DOI: 10.1080/19491247.2022.2037177.

Index

Printed in the United States
by Baker & Taylor Publisher Services